T0137205

Petroleum Engineering

The Springer series in Petroleum Engineering promotes and expedites the dissemination of new research results and tutorial views in the field of exploration and production. The series contains monographs, lecture notes, and edited volumes. The subject focus is on upstream petroleum engineering, and coverage extends to all theoretical and applied aspects of the field. Material on traditional drilling and more modern methods such as fracking is of interest, as are topics including but not limited to:

- Exploration
- Formation evaluation (well logging)
- Drilling
- Economics
- Reservoir simulation
- Reservoir engineering
- Well engineering
- Artificial lift systems
- Facilities engineering

Contributions to the series can be made by submitting a proposal to the responsible publisher, Anthony Doyle at anthony.doyle@springer.com or the Academic Series Editor, Dr. Gbenga Oluyemi g.f.oluyemi@rgu.ac.uk.

More information about this series at http://www.springer.com/series/15095

J. Carlos de Dios · Srikanta Mishra ·
Flavio Poletto · Alberto Ramos
Editors

CO_2 Injection in the Network of Carbonate Fractures

Editors
J. Carlos de Dios
CO2 Geological Storage
CIUDEN
Cubillos del Sil, León, Spain

Flavio Poletto
Geophysical Section
OGS
Trieste, Italy

Srikanta Mishra
Energy and Environment Business Unit
BATTELLE
Columbus, OH, USA

Alberto Ramos
Engineering School of Mines and Energy
UPM
Madrid, Spain

ISSN 2366-2646 ISSN 2366-2654 (electronic)
Petroleum Engineering
ISBN 978-3-030-62988-5 ISBN 978-3-030-62986-1 (eBook)
https://doi.org/10.1007/978-3-030-62986-1

This Springer imprint is published by the registered company Springer Nature Switzerland AG
The registered company address is: Gewerbestrasse 11, 6330 Cham, Switzerland

Contents

Light Drilling, Well Completion and Deep Monitoring

J. Carlos de Dios, Juan A. Marín, Carlos Martínez, and Alberto Ramos

Abstract Drilling is within the core activities of CO_2 geological storage since the more wells are drilled the higher amount of data is managed for site characterization and for a successful decision making on project viability. Most of commercial projects worldwide are at the early stage where the costs related to exploration play a key role, as is the case of the traditional drilling techniques from Oil and Gas industry that are usually expensive for on-shore projects whose business model still has to be proven. How to save drilling costs is addressed in the chapter, showing the experiences gained during the construction of the on-shore pilot: Hontomín Technology Development Plant (Burgos, Spain). Hontomín well drilling/completion was a success as the depth of 1600 m was reached using light drilling rigs (mining technique), achieving cost saving close to 60% in comparison to traditional techniques. Some experiences exist in the use of these rigs for mining, shale gas and oil and geothermal recovery, but for CO_2 geological storage they are limited to the Hontomín case. The existing technological drilling gaps identified during the plant construction and the future works for improving these rigs to reach the depth of 2500 m with a well geometry adequate to install advanced monitoring, are also addressed in this chapter.

Keywords On–shore CO_2 geological storage · Light drilling · Hontomín TDP · Cost savings · Target: 1600/2500 m depth · Advanced monitoring

J. C. de Dios (✉) · J. A. Marín
Foundation Ciudad de la Energía-CIUDEN F.S.P, Avenida del Presidente Rodríguez Zapatero, 24492 Cubillos del Sil, Spain
e-mail: jcdediosgonzalez@gmail.com; jc.dedios@ciuden.es

C. Martínez · A. Ramos
School of Mines and Energy, Technical University of Madrid, Calle de Rios Rosas 21, 28003 Madrid, Spain

J. C. de Dios et al. (eds.), *CO_2 Injection in the Network of Carbonate Fractures*, Petroleum Engineering, https://doi.org/10.1007/978-3-030-62986-1_1

1 Introduction

Well drilling is a core activity in the site characterization for CO_2 geological storage, and undoubtedly the more costly exploration work that impacts directly the project viability [1]. Therefore, the right election of drilling technique and equipment to use plays a key role that conditions the project as a whole.

For on-shore sites [2], the traditional techniques from Oil and Gas industry are usually expensive for an activity whose business model is still to be proven. The use of light equipment for drilling adapted from the mining industry offers the ability to achieve fully cored and completed wells with significant cost savings in comparison to the petroleum techniques [3]. Some experiences exist in the use of these rigs for mining exploration, shale gas and oil and geothermal recovery, but for CO_2 geological storage they are limited to Hontomín Technology Development Plant (TDP) case [4].

Hontomín is the unique on-shore injection site in Europe, located close to the city of Burgos in the north of Spain and operated by Foundation Ciudad de la Energía (CIUDEN). It has been recognized by the European Parliament as a "key test facility" to move forward the CCUS technologies to become a proven mitigation tool for the harmful effects produced by the emissions of greenhouse gases that cause climate change [5].

Two wells were drilled at Hontomín using mining technique to reach 1600 m depth with adequate geometry dimensions of completion and planned well trajectory, one intended for injection and the other for observation. The original plan was to use conventional rigs from Oil and Gas industry, but finally the light equipment was selected which meant cost saving of up to 60%. This was undoubtedly the main challenge to overcome during Hontomín pilot construction.

The use of light drilling rigs at Hontomín TDP construction was useful as wells were completed and monitored according to the planned design. This technology not previously used, also allowed relevant cost savings in comparison to traditional petroleum techniques as mentioned above. Undoubtedly these achievements are among the most relevant of Project "Compostilla OXYCFB300" [6], and they lead to think that light equipment can be used to reach depths of up to 2500 m with a well geometry adequate to install advanced monitoring, improving the effectiveness of the traditional drilling techniques.

However, relevant technological gaps were detected during well drilling at Hontomín related to safety and efficiency of the works performed. Preliminary studies have been carried out in ENOS Project [3] to analyze industrial solutions for improving the works conducted by these rigs.

Drilling techniques used at Hontomín, well completion and deep monitoring are addressed in this chapter, tackling the efficiency and safety of the works conducted on site, and the relevant cost savings that have been a success of Compostilla Project. Likewise, the current technological gaps associated with the use of these rigs are analyzed, as well as, the new technology development lines needed to improve light drilling technology.

2 Hontomin Well Drilling

Hontomín is a deep saline aquifer formed by naturally fractured carbonates. The main reservoir/seal pair is composed by Jurassic limestones and seal rocks belonging to the Lias and overlying Dogger formations, of which the primary hemipelagic seals are marls and black shales of Pliensbachian and Toarcian age. The site is a structural dome with the reservoir and seal rocks being located at a depth from 900 (top of the dome) to 1832 m (flanks). The main seal is the Marly Lias and Pozazal formations (highly carbonated marls, 160 m thick) and the reservoir is the Sopeña Formation (limestones and dolomites, 120 m thick) [7]. Both have a high level of fracturing in different geological blocks, but this does not affect the seal integrity.

The seal is formed by rock massifs with high uniaxial strength values close to 130 MPa and Young modulus values between 15 and 30 GPa. These data reveal that it is a hard rock with elastic-plastic deformation. The limestones and dolomites that compose the upper and down parts of the reservoir show uniaxial strength values between 180 and 190 MPa and Young modulus values in the ranges of 30–60 GPa and 60–80 GPa respectively. Hence, they are rigid rocks with brittle behavior which justifies the existence of fractures in the Sopeña Formation.

Figure 1 shows Hontomín geological column with main formations of seal-reservoir pair.

2.1 *Goals and Constrains*

The target formations to reach by drilling were Marly Liassic and Sopeña, main cap rock and reservoir respectively, which are located in the depth range of 1270–1550 m at Hontomín site. Two wells were designed with the completion and deep monitoring shown in Fig. 2.

The injection well (HI) is used to pump CO_2, brine and other fluids from surface to the reservoir in order to assess the fluid transmissivity in fractured carbonates, and the evolution of reservoir parameters as bottom hole pressure (BHP), temperature (BHT) and gas saturation. Hence, the following well completion and monitoring devices were installed in the well: super duplex tubing anchored to the liner by a hydraulic packer (1433 m MD), two P/T sensors located below the packer, one Distributed Temperature Sensing system (DTS) and one Distributed Acoustic Sensing system (DAS) joined along the tubing, six ERT electrodes and one U-tube device for fluid sampling from the bottom hole.

CO_2 plume tracking and other reservoir fluids evolution are monitored in the observation well (HA) that is equipped with fiberglass tubing anchored to the liner with 3 inflatable packer (1.275 m, 1.379 m and 1.479 m MD) that distribute the open hole in intervals with different permeability, four pressure/temperature (P/T) sensors and 28 ERT electrodes installed in the seal and reservoir formations.

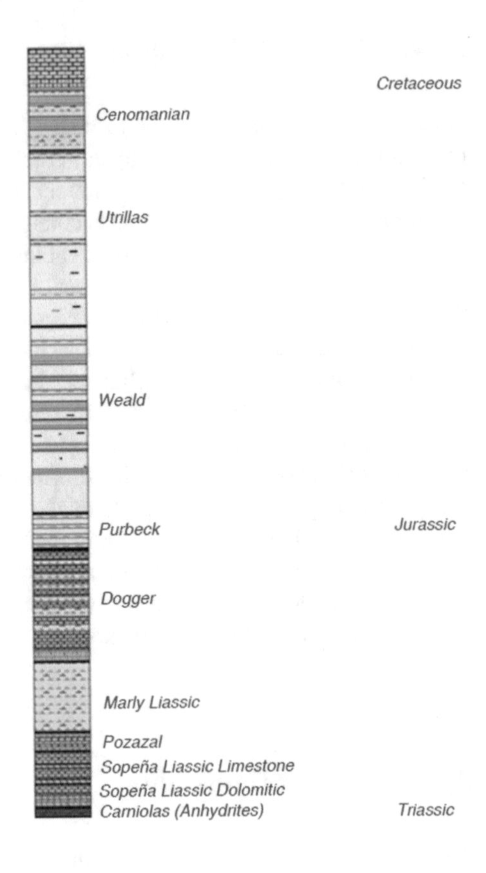

Fig. 1 Hontomín geological column

As mentioned above, the main challenge to overcome was to reach the depth of 1600 m according to the well geometry designed to install deep monitoring devices, using light drilling rigs. Never before these rigs had been used to do a work as planned at Hontomín. So, doubts raised previously the work startup like if finally the rigs would be able to reach the reservoir bottom, and in that case, if the well inner space to install the monitoring devices would be enough, and particularly, how the work efficiency would be and if drilling could be conducted accordingly existing safety standards. Relevant collaboration efforts were necessary between the drilling company staff/crew and CIUDEN engineering team to overcome daily problems during well drilling.

2.2 Drilling Rigs

Well drilling was performed at Hontomín with two light rigs: SEGOQUI 1900 and SEGOQUI 2000 (Fig. 3) using mining technique. SEGOQUI 1900 drilled first 600 m

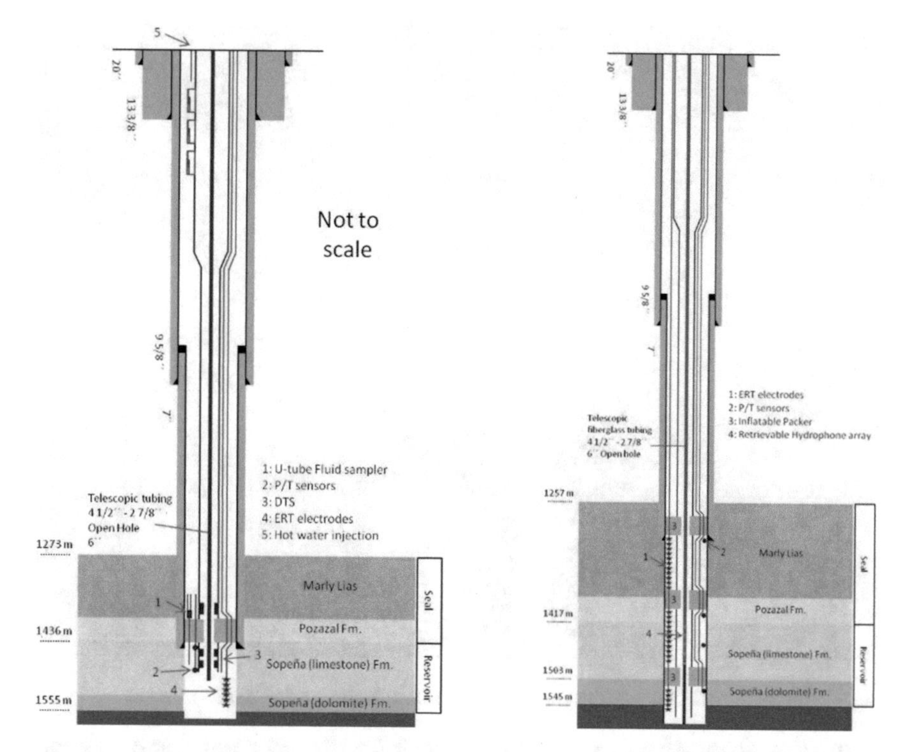

Fig. 2 Well completion/monitoring schemes of injection (left side) and observation (right side) wells of Hontomín TDP

depth and SEGOQUI 2000 was used to reach the target of 1600 m depth. Main technical characteristics of SEGOQUI 2000 are the following [8]:

- Mast height: 15.5 m.
- Engine power: 300 HP.
- Maximum torque: 4000 kg m.
- Rotary table opening: 150 mm.
- Cylinder hoisting load: 50 t.
- Winch load: 60 t.
- Total load (cylinder + winch): 110 t.
- Maximum push load: 20 t.
- Maximum speed: 120 rpm.
- Drill pipe: φ 140, 152 mm L 6 m.
- Rig mounted on truck 8 × 8.

Following auxiliary equipment and infrastructure were also necessary (Fig. 4):

- 2 Compressors Atlas Copco XRVS 455, 25 bar and 25 m^3/min.
- 1 Booster HURRICANE M 41C-870, 60 bar and 50 m^3/min.
- Mud pump GARDNER-DENVER Mod 7 1/4″ × 14″ × 10″ and 5″ × 10″.

Fig. 3 Rig SEGOQUI 2000 drilling at Hontomín TDP (Courtesy of CIUDEN)

Fig. 4 Drilling on-site panoramic view with the rig and auxiliary equipment (Courtesy of CIUDEN)

- Mud pump EMSCO F-500. Triplex Mud Pumps (API-7K) 500 HP.
- Screen and double cyclone MODELCO model MD 190 D 200 m^3/h.
- 2 mud pools. Total capacity 75 m^3.
- Electricity generator: 25 kVA for lighting.
- Mud logging cabin.
- Geological control cabin, equipped with chromatograph for gases and masterlog software.
- H_2S and CH_4 continuous monitoring.
- BOP (Blowout Preventer): WP 5000 psi.
- Choke manifold and torch.
- Crane and auxiliary vehicles.
- Pipe for direct and reverse drilling.
- Core sampling pipe and bits (OD 80 mm and 7 m length). OD 6″ cores.

2.3 *Workflow*

Drilling process for both injection and observation wells was as follows:

1. Percussion drilling for first 130 m depth.
2. Rotary drilling by reverse mud circulation up to reach the bottom of Utrillas Formation (600 m depth).
3. Rotary drilling by direct mud circulation up to reach the top of Keuper Formation (close to 1600 m depth).

Percussion drilling was performed using a trepan for first 130 m, in order to ensure the well alignment and verticality. Afterwards, considering that first shallow relevant formation crossed was Utrillas (see Fig. 1), which is comprised of sand, gravel and little cohesive material in general terms, the reverse mud circulation drilling was conducted due to the good performance of this technique for this ground. Finally, direct mud circulation drilling was used from 600 to 1600 m depth, crossing the upper seal formations: Weald, Purbeck, Dogger, Marly Liassic and Pozazal, the reservoir Sopeña Formation and finally reaching the Carniolas (anhydrites) at the top of Keuper (see Fig. 1).

Reverse mud circulation technique [9] was performed insufflating compressed air in the inner part of the drill pipe OD 220 mm and 6 m length, through a valve installed to the depth of between 200 and 250 m. The compressor Atlas Copco XRVS 455, 25 bar and 25 m^3/min was used for this maneuver. Below the valve, standard drill pipe 5 1/2″, 9–10″ loading bars and 17 1/2″ or 12 1/4″ bits were installed. This maneuver produces the air-lift effect, lightening the hydrostatic column above the valve and inducing the extraction of the mud along the inner part of drill pipe. The process scheme for reverse circulation is shown in Fig. 5.

A Blowout preventer valve (BOP) [10] 11″ 5000 psi was installed previously to start 3rd drilling phase (exploration of cap-rock and reservoir formations) for avoiding the risk of gas eruption from the bottom hole, as shown in Fig. 6. The

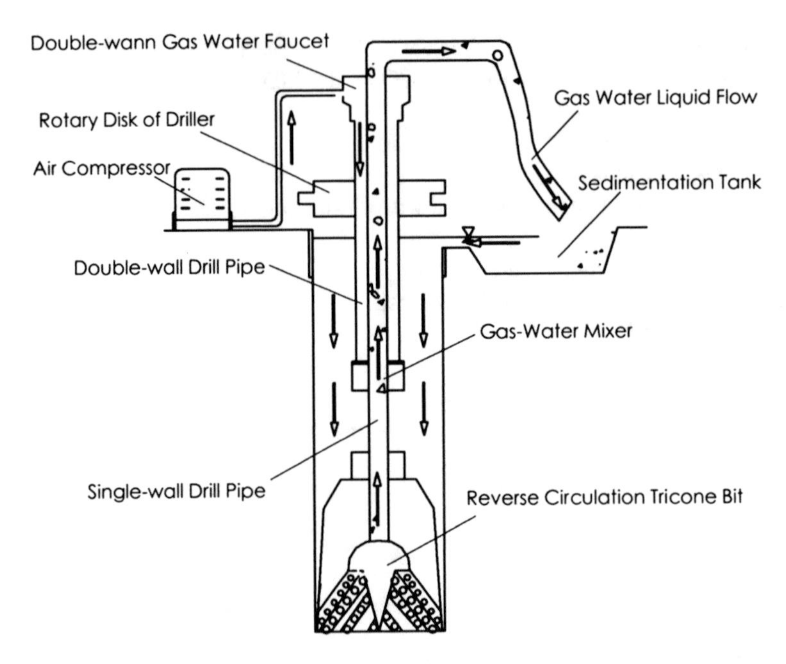

Fig. 5 Reverse mud circulation drilling scheme

Fig. 6 Installation of BOP and spools during HI well drilling (Courtesy of CIUDEN)

spools, pressure lines and BOP were tested by hydraulic pressure of 70 bar held constant for 15 min. No variations were detected during the period. Subsequently, direct circulation drilling started for the length range 600–1600 m depth.

As mentioned above, direct mud circulation drilling [11] was used to cross the seal (Marly Lias and Pozazal) and reservoir (Sopeña). It was conducted pumping the mud through the inner part of drill pipes to reach the bottom hole, producing the bit cooling and lifting the rock cuttings to the surface. The cake around the wellbore

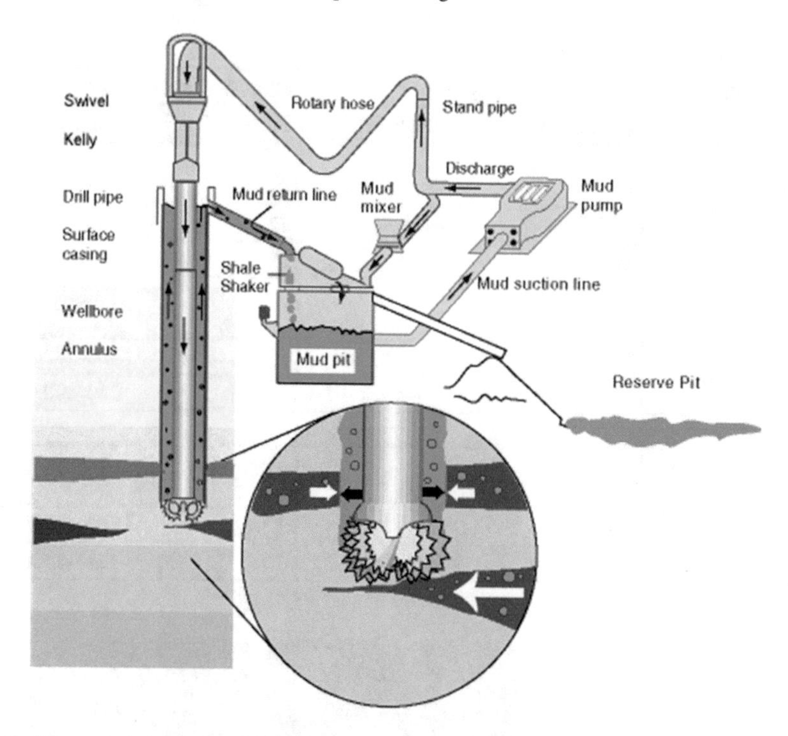

Fig. 7 Direct mud circulation drilling scheme (Courtesy of Massenza drilling rigs)

was also built during the maneuver to avoid the collapse of wellbore wall and acting as a first barrier in gas release case, as shown in Fig. 7.

Table 1 shows drilling bit diameters used and depth reached in each process phase (Fig. 8).

The drilling mud was made with the following components:

- Bentonite to increase the density and viscosity of the fluid.
- Calcium carbonate to increase the density.
- Carboxy Methyl Cellulose for filtering and viscosity control.
- Sodium Hydroxide to control the alkali and pH.
- Sodium carbonate for density control.
- Dry polymer to reduce the friction.
- Agent for rheological control.
- Antifoaming.

Its main characteristics are:

- Density: 1.01–1.09 g/cm^3.
- Funnel Viscosity: 41–48 s.
- Apparent Viscosity: 14–21 cp.
- Yield Point: 10–18 lb/100 ft^2.

Table 1 Drilling phase, bit OD and depths

Drilling phase	Injection well (HI)		Observation well (HA)	
	Bit OD (″)	Length MD (m)	Bit OD (″)	Length MD (m)
1st phase: percussion				
Percussion 1	25 1/4	0–20	22	0–20
Percussion 2	19	20–130	19	20–132
2nd phase: reverse mud circulation				
Reverse 1	17 1/2	130–210	17 1/2	132–220
Reverse 2	12 1/4	210–591	12 1/4	220–600
BOP installation				
3rd phase: direct mud circulation				
Direct 1	8 1/2	591–1441	8 1/2	600–1286
Direct 2	6	1441–1570	6	1286–1580

Fig. 8 Drilling bits used at Hontomín (Courtesy of CIUDEN)

- pH: 9–10.
- Cake: 0.5 mm.

Mud circulation features were: flow rate between 1000 and 1200 l/m and pressure ranges of 5–10 bar and 15–25 bar for 2nd and 3rd drilling phases respectively.

2.4 Well Completion and Deep Monitoring

Figure 9 shows the following completion components (according to API standards) and monitoring devices installed in the injection and observation wells.

Completion

- **HI Well**:
  - 20″ conductor, S235JR (20 m depth), 13 3/8″ casing, 61 lb/ft, K55, BTC (207 m depth), 9 5/8″ casing, 43.5 lb/ft, N 80, BTC (586 m depth) and 7″ Liner, 29 lb/ft, N80, BTC (from 483 to 1437 m depth, last 200 m L80Cr13).
  - Tubing 4 1/2″, 13.5 lb/ft, CR22-140, VAM TOP, R2/R3 (408 m depth), tubing 2 7/8″, 7.8 lb/ft, 22 CR-125 (from 408 to 1466 m depth).
  - Tubing hanger (L = 0.76 m) (GL at bottom of the TH) and X-over 4 1/2″ EUE pin × 4 1/2″ EUE pin (L = 0.30 m).
  - 7″ RDH Dual Hydraulic-Set/Retrievable Packer, 5000 psi WP, 13Cr and Nitrile element, Primary connection 2 3/8″ API-EU box × 2 3/8″ API-EU pin + X-Over pin × pin (L = 2.80 m) (1431 m depth).
  - Otis 1.875″ X Selective Landing Nipple, X20Cr13, 2 3/8″ API EU pin × pin (L = 0.26 m) and choke (1003 m depth), Sliding Side-Door Circulating Device, 1.875″ X Profile, 13Cr, 2 3/8″ EUE pin × pin (L = 1.02 m) (1417 m depth), Otis 1.875″ X Selective Landing Nipple, X20Cr13, 2 3/8″ API EU pin × pin (L = 0.27 m) (1444 m depth), Otis 1.875″ XN Landing Nipple (Bottom No-Go), X20Cr13, 2 3/8″ API EU pin × pin (L = 0.31 m) (1456 m depth) and RH Catcher Sub Bell Type, X20Cr13, 2 3/8″ EUE box up (L = 0.15 m) (1466 m depth).
  - 3 Sidepocket Mandrel, 13Cr, with RD-2 Dummy Valve, 4 1/2″ EUE box × pin (L = 3.02 m) (176, 281 and 383 m depth).
- **HA Well**:
  - 20″ conductor, S235JR (20 m depth), 13 3/8″ casing, 61 lb/ft, K55, BTC (216 m depth), 9 5/8″ casing, 43.5 lb/ft, N 80, BTC (594 m depth) and 7″ Liner, 29 lb/ft, N80, BTC (from 490 to 1281 m depth, last 200 m L80Cr13).
  - 4 1/2″ tubing, 7.6 kg/m, EPOXY/FG, Serial number 2500, T&C EUE 8 RD (371 m depth), tubing 2 7/8″, EPOXY/FG, Serial number 2500, T&C EUE 8 RD (from 371 to 1561 m depth).
  - Tubing hanger (L = 0.76) (GL at bottom of the TH) and X-over 4 1/2″ pin × 4 1/2″ EUE box (L = 0.21 m).
  - 3 Inflatable Packers, SS 316L and HNBR Nitrile, 2 × 1/4″ infl/des. lines (2 7/8″ EUE box × box) (L = 2.79 m) (1275, 1380 and 1498 m depth).
  - 1.875″ XN Landing nipple (Bottom No-Go), 13 Cr, 2 7/8″ 6.5 API EUE (L = 0.45 m) (1508 m depth).

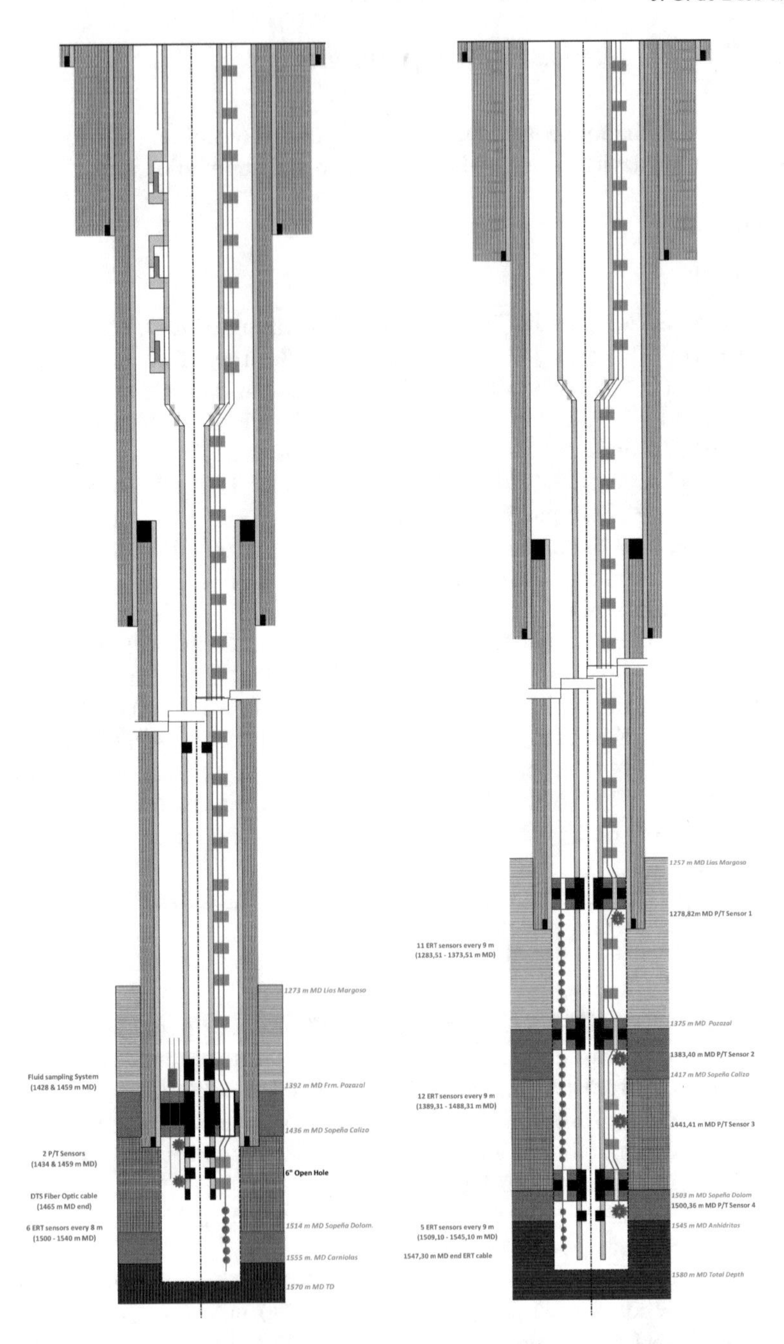

Fig. 9 Schemes of well completion and monitoring of the injection (on the left) and observation (on the right) well

Monitoring

- **HI Well**:
  - U Tube sampling system (1459 m depth).
  - Distributed Temperature Sensing System (DTS) (along the tubing, 1465 m depth).
  - Distributed Acoustic Sensing system (DAS) (along the tubing, 1465 m depth).
  - P/T sensors (1434 and 1459 m depth).
  - 6 ERT (from 1500 to 1540 m depth).
- **HA Well**:
  - 4 P/T sensors (1279, 1383, 1441 and 1500 m depth).
  - 11 ERT sensors (from 1283 to 1373 m depth).
  - 12 ERT sensors (from 1389 to 1488 m depth).
  - 5 ERT sensors (from 1509 to 1545 m depth).

Note. The data to locate the components of completion and monitoring are measured in depth units (m MD).

Casings and liners were cemented using CO_2 resistance cement for avoiding damages due to the acidification produced by the mixture of carbon dioxide and reservoir saline water. CBL (Cement Bond log) logging device was used for checking the cementing grade.

2.5 *Coring*

The extraction of rock samples during well drilling, known as coring, is a key activity to reduce uncertainty in the seal-reservoir evaluation by providing data representative in situ conditions [12]. Samples are used to perform laboratory scale tests to determine the reservoir injectivity, storage capacity and long term trapping [13].

Drilling and extraction of rock samples are part of the drilling report, as shown in Tables 2 and 3, which include the information of coring activity conducted during the injection and observation well drilling.

Coring was conducted by ID 6″ drilling tool and piping for sample recovery, as Fig. 10 shows.

Finally, 13 core samples were acquired from well drilling, 10 from the observation well (HA) and 3 from the injection well (HI), of which 7 correspond to the caprock (Marly Lias and Pozazal Formations) and 6 to reservoir (4 from Limestone and 2 from Dolomitic Sopeña Formations) (Fig. 11).

Petrophysical routine and specific lab tests [4, 13] were carried out using these rock cores to determine the ability of formations to be cap-rock and reservoir respectively. Lab procedure description and analysis of results to determine petrophysical properties, injectivity, impacts of hydrodinamic, mechanical and geochemical effects due to injection and trapping will be addressed in Chapter “Laboratory Scale Works”.

Table 2 Observation well (HA) coring information

Core sample	Recovery		Drilling interval (m)
	Length drilled (m)	%	
1	7	100	1307–1314
2	3.5	100	1320–1323.5
3	6	85.7	1343–1349
4	4	100	1401–1405
5	5.10	72.8	1405–1410
6	6.77	97	1442–1449
7	1.38	100	1449–1450.38
8	5.87	0	1457–1462.87
9	0.12	60	1464–1462.1
10	6.91	98.7	1515–1522

Table 3 Injection well (HI) coring information

Core sample	Recovery		Drilling interval (m)
	Length drilled (m)	%	
1	7	100	1355–1362
2	0.96	19	1467.74–1468.7
3	6.96	99.12	1531–1538

Fig. 10 Drilling tool and pipe for coring used at Hontomín (Courtesy of CIUDEN)

Fig. 11 Core sample from limestone Sopeña Formation in the observation well (Courtesy of CIUDEN)

2.6 *On-Site Tests*

Well logging works [14] conducted at Hontomín site were performed using petrophysical probes and other running tools along the completion and open hole of both wells, in order to achieve the following goals:

- Data acquiring for a better knowledge of geophysical characteristics and more accurate location of the pair seal-reservoir.
- Check well drilling geometric parameters as its azimuth, inner diameter and vertical deviation.
- Perform quality control of well completion, particularly the cementing grade and existence of leakages.

Well logging works focused to increase the knowledge of geological formations conducted within the site characterization were as the following:

- Gamma ray (natural gamma ray and gamma log) to locate the top and bottom of each geological formation, the clay amount existing in their composition and existing density.
- Temperature to determine the thermal gradient related to depth.
- Neutron to determine rock matrix porosity of each formation.

Spontaneous potential used promptly to identify the aquifer limits and brine movement direction.

- Resistivity to determine brine conductivity and its salinity.
- Sonic to determine the velocities Vp and Vs along the open hole and their correlation with geomechanical properties.
- Acoustic televiewer to identify the azimuth and dip of main fractures existing in the borehole.

Particularly, well logging works conducted to check well completion quality were the following:

- Caliper used to determine bottom-hole inner diameter.
- Gyroscope used to determine the azimuth and deviation of well drilling path.
- CBL (Cement Bond Log) used to determine cementing level of casings and liners.

Figure 12 shows results from several well logging probes run in the observation well (HA), plotted in the same scheme for its correlation and interpretation. This work was conducted in the depth range from 1280 to 1570 m combining the following tools:

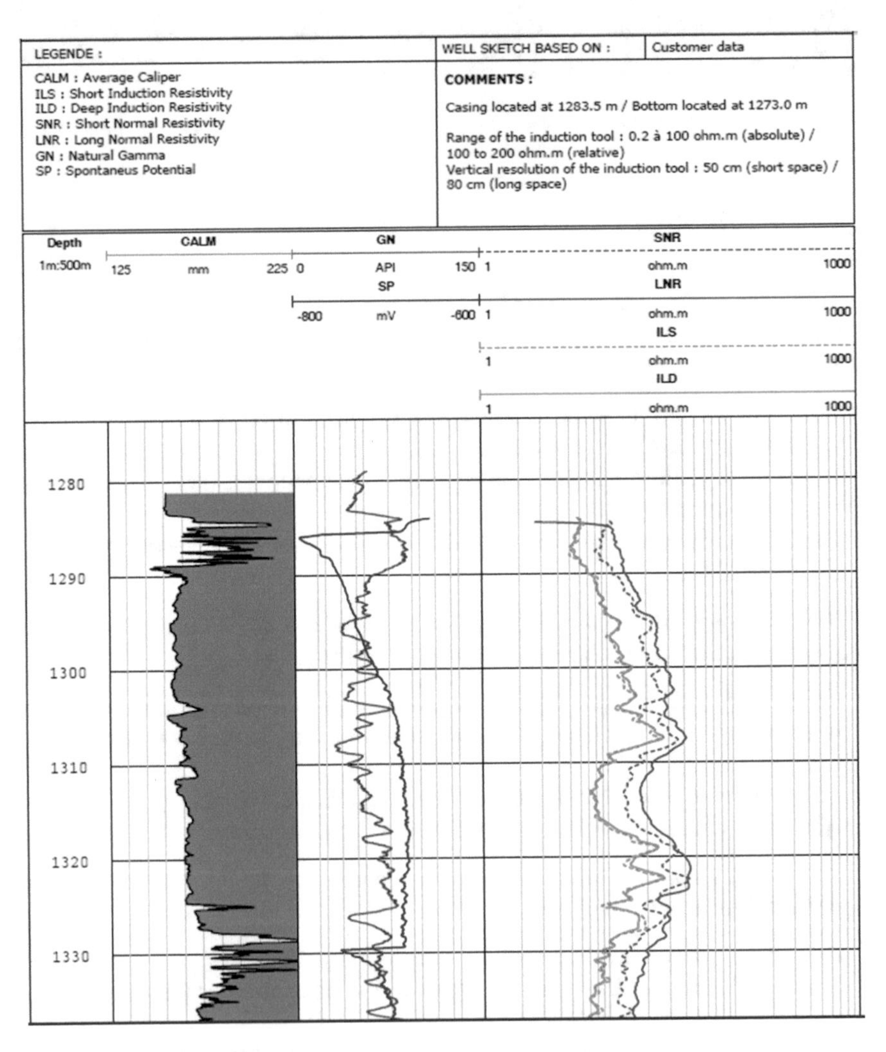

Fig. 12 Observation well logging results

- Caliper
- Short and deep induction resistivity
- Short and long normal resistivity
- Gamma ray
- Spontaneous potential.

Other logging probe run at Hontomín wells was the acoustic televiewer [4, 15] used to give an oriented and continuous borehole-wall imaging for fractured-rock aquifer studies. This type of well logging will be addressed in Chapter "On-Site Hydraulic Characterization Tests". Finally, following on-site tests were performed during site hydraulic characterization [4], which will be also described and analyzed in Chapter "On-Site Hydraulic Characterization Tests":

- Permeability test at field scale (PTFS)
- Connectivity test inter wells (CTIW)
- Leak off test (LoT).

3 Performance Curves and Cost Saving

Drilling performance curve (DPC) [16] is the tool for assessing the operation performance for a specific well or for series drilled in an area with same technology and workflow. The graphic provides the information needed to analyze the working sequence in the well(s) and the time taken to reach a given depth.

Figure 13 shows the HI/HA Well DPCs which include planned works (in red line) and real execution (in blue line) of following activities:

- Percussion drilling.
- Reverse circulation rotary drilling.
- 1st completion.
- Direct circulation rotary drilling.
- 2nd completion.
- Well logging.
- Coring.
- Downtimes.

Tables 4 and 5 shows the correlation between each activity and time required during the drilling of both wells.

Although drilling performance is lower than case where oil and gas rigs are used [17], well light drilling was a success as the depth of 1600 m was reached at Hontomín, which had not been done before, achieving cost savings close to 60% in comparison to traditional techniques. Main reason is the associated costs to light drilling are considerably lower than those corresponding to Oil and Gas drilling, since both the availability, transport and operation of light rigs is less expensive. Nevertheless, several technological gaps were identified during Hontomín construction, as well as the necessity to find solutions that make more reliable operations, increasing the

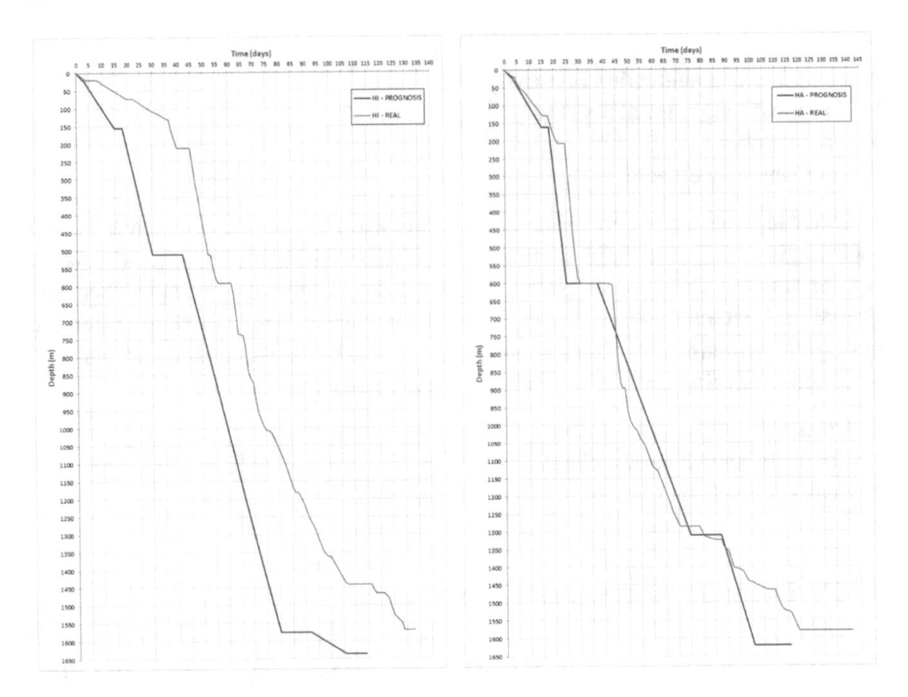

Fig. 13 HI/HA drilling performance curves

Table 4 HI well drilling and completion, associated works and time

HI well activities	Time (days)	Well drilling average (%)
Drilling	74	48.88
Maneuvers	17	10.87
Equipment maintenance	1	0.79
Circulation/losses/mud manufacturing	4	2.44
Piping placement/cementing	13	8.88
Coring	6	4.2
Well logging	10	6.84
Staff rest	3	1.72
On-site tests	10	6.14
Final well completion (monitoring, well heads)	12	7.78
Downtimes (e.g. tool fishing, fault repair)	2	1.46
Total	152	100

Table 5 HA well drilling and completion, associated works and time

HA well activities	Time (days)	Well drilling average (%)
Drilling	54	33.17
Maneuvers	19	11.8
Equipment maintenance	2	0.94
Circulation/losses/mud manufacturing	3	1.9
Piping placement/cementing	6	3.91
Coring	18	10.68
Well logging	8	4.95
Staff rest	9	5.68
On-site tests	22	13.2
Final well completion (monitoring, well heads)	9	5.52
Downtimes (e.g. tool fishing, fault repair)	13	8.25
Total	163	100

safety and efficiency. Section 4 address this issue, classifying the gaps according to their impacts on drilling workflow and on work efficiency and safety.

4 Existing Technological Gaps

Technological gaps identified during Hontomín well drilling are related to the following topics:

- Operation efficiency and safety
- Well completion and the installation of monitoring devices
- Rig instrumentation and operation control
- Directed drilling.

The gap analysis described below will be used in future works as a reference to explore the existence of technological solutions.

4.1 Gaps Related to Operation Efficiency and Safety

Regarding well drilling efficiency, first general matter to analyze is whether it is possible that an unique light rig is able to develop the works performed in Hontomín by the following equipment:

- Percussion drilling rig for first 130 m depth.
- Rotary drilling with reverse mud circulation up to reach 600 m depth.
- Rotary drilling direct mud circulation up to reach 1600 m depth.

If so, downtimes due to rig placement and disassembly would be avoided.

On the other hand, having a look to Tables 4 and 5, drilling performance is a crucial parameter as it was in the range 33.17–48.88% of total well drilling period. Therefore, parameters such as the push, rotary speed, torque and pull, and related operations as mud pumping, must be analyzed for their improvement. Likewise, the drill pipe placement and disassembly must avoid as much as possible downtimes, so that, pull capacity and rig height play a key role. On one hand, pull capacity provides high lift velocity and the ability to support heavy loads at deep depths. On the other, higher rig height allows the use of longer drill pipe and install longer tubing parts than used at Hontomín works, which means a performance increase in piping handling operations. These improvements could reduce the maneuver time that corresponds to 11% of well drilling period.

Regarding safety issues, the capability to integrate the additional equipment (Fig. 4) in the rig is really important, taking into account the more components are needed to be installed and dissembled the higher probability of accident occurs. It is particularly relevant that the rig can assembles the blow-out preventer valve, for which, the elevation of the machine and a specific mechanical and hydraulic coupling design are needed. In the same way, an integrated equipment is more efficient since it is needed less time to install and disassembly components.

4.2 Gaps Related to Well Completion and the Installation of Monitoring Devices

Drilled diameter is a critical parameter because of the necessity to install deep monitoring devices inside the annular space between the casing/liner and tubing in some cases, and within the interphase between the rock and the external part of well completion in others.

Likewise, well diameter conditions tubing dimensions, being this fact more critical in the injection well case, since a nominal flow rate is planned and the diameter value is conditioned by the steady state injection in laminar flux. In any case, drilling diameter increase depends on rig capability determined by the parameters described above, mainly the push, rotary speed, torque, pull and mud pumping.

For industrial injections, tubing OD is planned to be equal or higher than 4–5 1/2″ that corresponds to 9 5/8″ OD casing/liner and 12 1/4″ drilled diameter. If well monitoring devices are decided to assembly in the outer completion part instead of the inner annular, drilling diameter should be increased. Diameters for Hontomín bottom holes were 8 1/2″ and 6″ in the completed and "open hole" areas respectively. This fact conditioned both the final tubing dimension (2 7/8″) and the installation process of monitoring devices, which was a risky operation due to the tight annular space existing between the tubing and the borehole wall.

On the other hand, the reason why cementing is crucial in drilling for CO_2 geological storage is the wells are main potential migration pathways if cementing grade is not as required. Besides this critical point, cementing is usually a well service provided by an external company. This fact involves the adversities of a contract, what supposes high budgetary conditions in this case and lack of immediate quality control if well logging is also performed by an external service. Well repair in case of faulty work is really difficult and costly. These reasons lead to thinking that cementing could be included as other activity to be conducted by own drilling staff.

4.3 *Gaps Related to Rig Instrumentation and Operation Control*

The instrumentation of rigs used in Hontomín well drilling corresponds to usual equipment for shallow operations, as mining exploration and hydrogeological prospecting, being 1000 m depth the common barrier for this type of works. Therefore, the rig instrumentation must be improved to reach depths in the range 1600–2500 m in an efficient and safe manner.

Pumping facility and its control were not integrated as part of the rig, what means operation difficulties that produce inefficiency and working faults may occur. Hence, it would be advisable to analyze how integrate the mud facility on drilling equipment, and particularly, its operation control.

Regarding data of Tables 4 and 5, almost 5–7% of drilling period was used to conduct well logging works. These ones were performed by an external service company, what meant extra costs and coordination efforts regarding the availability to conduct the planned works. Particularly, borehole cross-section diameter is usually controlled by the caliper log, a running tool which determines the cross section deviation from planned for different well parts. Graphic of Fig. 14 shows the caliper logging results.

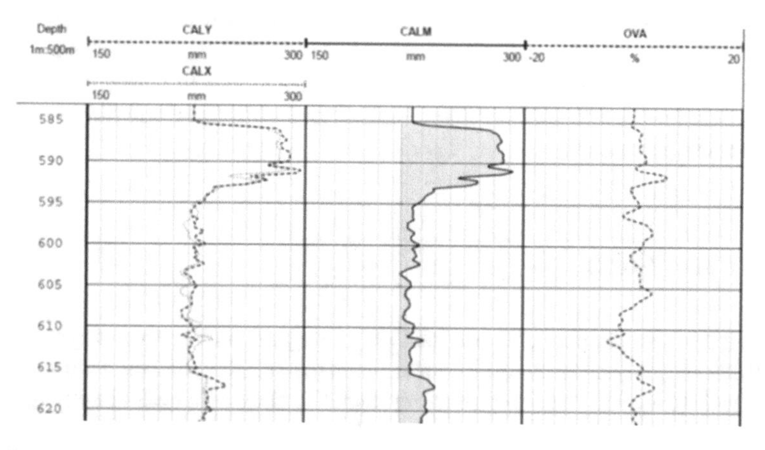

Fig. 14 Caliper logging results

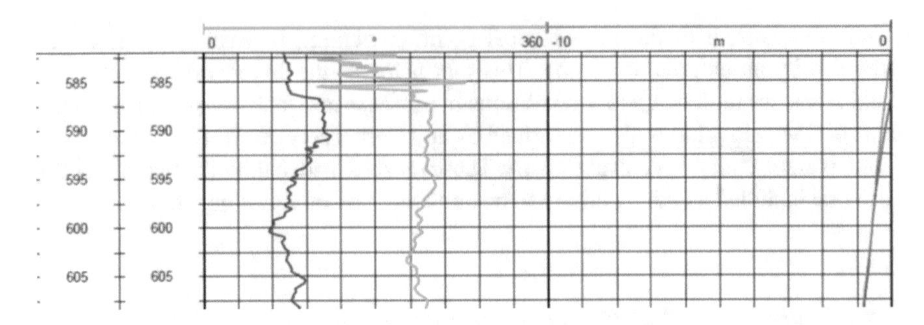

Fig. 15 Gyroscope logging results

In a similar way, borehole deviation control is developed by the gyroscope logging, a running tool which determines the well axis deviation from planned for different intervals in depth. Figure 15 shows the gyroscope logging results.

Clearly, if data logging can be conducted during drilling, less downtimes will occur, as drill pipe is located along the well and no need of maneuvers for place and replace. On the other hand, a quick information on drilling performance may be achieved which is crucial to take the right decision at the right time. Therefore, it is necessary to determine what type of monitoring devices would be suitable to install within the drilling array, in order to carry out logging and measuring while drilling (L/MWD), particularly to control the borehole cross-section and deviation [18, 19].

4.4 Directed Drilling

Directed drilling is more and more required. Initially, traditional Oil and Gas industry decided to use this type of wells for particular reasons, and mainly for unconventional hydrocarbon recovery techniques. More recently other activities as geothermal, energy storage and CO_2 geological storage have claimed the use of this technique. Unfortunately, the existing light drilling rigs are not able to perform this type of works, being needed to analyze technological solutions to perform directed drilling using these equipments.

5 Future Works

Drilling is within the core activities of CO_2 geological storage since the more wells are drilled the higher amount of data is managed for site characterization and for a successful decision making on project viability, since it is undoubtedly the more costly exploration work.

For this reason, a feasibility study to improve light drilling rigs used at Hontomín was carried out in ENOS Project [3], whose goals in this matter are the following:

- Reach depth of 2500 m at least
- Achieve enough inner space at well bottom hole to install monitoring devices.

Ciudad de la Energía Foundation (CIUDEN), Bureau de Recherches Geologiques et Minieres (BRGM) and Sotacarbo, as project partners, have counted on the technical collaboration and advice of HERRENKNECTH AG, company specialized in tailor made manufacturing of deep drilling rigs, that supports and promotes their use for geological exploration related with future energy needs.

6 Concluding Remarks

Main concluding remarks on light rigs use for deep well drilling regarding the experience gained during the construction of Hontomín Technology Development Plant for CO_2 geological storage are the following:

- Although light drilling had not previously been used in wells more than 1000 m depth, this was a success during Hontomín pilot construction.
- Final depth close to 1600 m was reached with well dimensions appropriate to install deep monitoring devices.
- Cost savings were up to 60% in comparison to those corresponding to traditional oil and gas techniques.
- The achievements described above lead to thinking that light drilling technique may be used to reach the depth of 2500 m with a well geometry adequate to install advanced monitoring.
- Drilling equipment improvement is necessary to achieve mentioned goals, since technological gaps that impact on drilling operations and on work efficiency and safety were identified during the pilot construction.

Acknowledgements The experiences and results showed in this chapter form part of the project "OXYCFB 300" funded by the European Energy Program for Recovery (EEPR) and the Spanish Government through Foundation Ciudad de la Energía-CIUDEN F.S.P. Authors acknowledge the role of the funding entities, project partners and collaborators without which the project would not have been completed successfully.

This document reflects only the authors' view and that European Commission or Spanish Government are not liable for any use that may be made of the information contained therein.

Glossary

API American Petroleum Institute
BHP Bottom hole pressure

BHT Bottom hole temperature
BOP Blowout preventer
CBL Cement Bond Log
CCUS Carbon capture utilization and storage
CTIW Connectivity test inter wells
DAS Distributed acoustic sensing system
DPC Drilling performance curve
DTS Distributed temperature sensing system
ENOS Enabling on shore CO_2 storage in Europe
EP Resolution European Parliament Resolution
ERT Electrical resistivity tomography
HA Hontomín observation well
HI Hontomín injection well
LoT Leak off test
LWD Logging while drilling
MD Measured depth
MWD Measured while drilling
OD Outer diameter
P/T Pressure/Temperature
PTFS Permeability test at field scale
TDP Technology Development Plant

References

1. Fukaia, I., Mishraa, S., & Pasumartia, A. (2017). Technical and economic performance metrics for CCUS projects: Example from the East Canton Consolidated Oil Field, Ohio, USA. *Energy Procedia, 114*, 6968–6979. Available online https://doi.org/10.1016/j.egypro.2017.03.1838. Accessed October 2, 2019.
2. Gastine, M., Berenblyum, R., Czernichowski-Lauriol, I., de Dios, J. C., Audigane, P., Hladik, V., et al. (2017). *Enabling onshore CO_2storage in Europe: Fostering international cooperation around pilot and test sites. Energy Procedia, 114*, 5905–5915.
3. *ENOS «Enabling onshore CO_2storage in Europe»*. Available online https://www.enos-project.eu/. Accessed October 2, 2019.
4. de Dios, J. C., Delgado, M. A., Marín, J. A., Salvador, I., Álvarez, I., Martinez, C., & Ramos, A. (2017). Hydraulic characterization of fractured carbonates for CO_2 geological storage: Experiences and lessons learned in Hontomín Technology Development Plant. *International Journal of Greenhouse Gas Control, 58C*, 185–200.
5. *European Parliament Resolution of 14 January 2014 on implementation report 2013: Developing and applying carbon capture and storage technology in Europe (2013/2079(INI)). Bullet 17.*
6. *The Compostilla Project «OXYCFB300» Carbon capture and storage demonstration project* (Knowledge sharing FEED report). Global CCS Institute Publications. Available online https://hub.globalccsinstitute.com/sites/default/files/publications/137158/Compostilla-project-OXYCFB300-carbon-capture-storage-demonstration-project-knowledge-sharing-FEED-report.pdf. Accessed October 7, 2019.

7. Rubio, F. M., Garcia, J., Ayala, C., Rey, C., García Lobón, J. L., Ortiz, G., & de Dios, J. C. (2014). Gravimetric characterization of the geological structure of Hontomín. In *8ªAsamblea Hispano-Lusa de Geodesia y Geofísica*, Évora.
8. *Talleres Segovia drilling experts*. Available online https://www.talleresegovia.com/es/drilling_experts/noticia/novedades/equipos_perforadores_segoqui_para_proyecto_de_almacenamiento_de_co2_cudein. Accessed October 9, 2019.
9. Graham, R. L., Foster, J. M., Amick, P. C., & Shaw, J. S. (1993). Reverse circulation air drilling can reduce well bore damage. *Oil and Gas Journal (United States), 91*, 12. ISSN 0030-1388. Available online https://www.osti.gov/biblio/6558198. Accessed October 7, 2019.
10. Pinker, R. (2012). *Improved methods for reliability assessments of safety-critical systems: An application example for BOP systems* (Master thesis). NTNU, Institutt for produksjons- og kvalitetsteknikk. Available online https://ntnuopen.ntnu.no/ntnu-xmlui/handle/11250/240823. Accessed October 10, 2019.
11. Australian Drilling Industry Training Committee Limited. (2015). *The drilling manual* (5th ed.). ISBN 9781439814208/1439814201.
12. Al-Saddique, M. A., Hamada, G. M., & Al-Awad, M. (2000). State of the art: Review of coring and core analysis technology. *Journal of King Saud University-Engineering Sciences, 12*(1), 117–137. Available online https://doi.org/10.1016/S1018-3639(18)30709-8. Accessed October 14, 2019.
13. de Dios, J. C., Delgado, M. A., & Álvarez, I. (2018). *Laboratory procedures for petrophysical characterization and monitoring of CO_2 geological storage in deep saline aquifers*. PTECO2 Publications. Available online https://www.pteco2.es/es/publicaciones/procedimientos-de-laboratorio-para-la-caracterizacion-petrofisica-y-el-control-de-almacenes-geologicos-de-co2-en-acuiferos-salinos-profundos
14. Simpson, D. A. (2018). *Practical onshore gas field engineering*. Elsevier, Gulf Professional Printing. ISBN 978-0-12-813022-3.
15. Williams, J. H., & Johnson, C. D. (2004). Acoustic and optical borehole-wall imaging for fractured-rock aquifer studies. *Journal of Applied Geophysics, 55*(1–2), 151–159. Available online https://doi.org/10.1016/j.jappgeo.2003.06.009. Accessed October 15, 2019.
16. Brett, J. F., & Millheim, K. K. (1986). The drilling performance curve: A yardstick for judging drilling performance. In *SPE Annual Technical Conference and Exhibition*, New Orleans, LA, October 5–8. Society of Petroleum Engineers. Available online https://doi.org/10.2118/15362-MS. Accessed October 17, 2019.
17. Cochener, J. (2010). *Quantifying drilling efficiency*. US EIA Department of Energy. (202) 586-9882. Available online https://www.eia.gov/workingpapers/pdf/drilling_efficiency.pdf. Accessed October 17, 2019.
18. Glover, P. (2016). *Caliper logs. Petrophysics MSc course notes*. Leeds University. Available online https://homepages.see.leeds.ac.uk/~earpwjg/PG_EN/CD%20Contents/GGL-66565%20Petrophysics%20English/Chapter%209.PDF. Accessed October 22, 2019.
19. Weston, J. L., Ledroz, A. G., & Ekseth, R. (2014). New gyro while drilling technology delivers accurate azimuth and real-time quality control for all well trajectories. *Society of Petroleum Engineers. SPE Drilling & Completion, 29*(3). Available online https://doi.org/10.2118/168052-PA. Accessed October 22, 2019.

Well Logging in Fractured Media

Autumn Haagsma, Isis Fukai, Erica Howat, Jared Schuetter, Amber Conner, Ben Grove, Srikanta Mishra, and Neeraj Gupta

Abstract This chapter provides an overview of well logging in carbonate formations from the perspective of CO_2 sequestration projects. The chapter begins with a summary of the various types of logs such as gamma ray, neutron porosity, bulk density, resistivity, photoelectric, sonic, pulsed neutron capture, nuclear magnetic resonance, image log, and elemental spectroscopy. Next comes a discussion of analyses performed with these log outputs for characterizing attributes such as: facies, shale volume, porosity, permeability, fluid saturation, geomechanical properties, and fractures and vugs. This is followed by best practices for log interpretation, including data QA/QC and a typical workflow. A case study on well-log interpretation is then presented using two fields from the Northern Michigan Pinnacle reef trend. This example covers data acquisition, porosity analysis, integration of core and log data, facies analysis, and fractures and vugs. The application of machine learning to well logging data is presented next, beginning with a discussion of machine learning basics, which is followed by an example application for vug characterization. Collectively, these topics cover the most common set of conventional and emerging techniques that are relevant in the utilization of well logs for characterization of carbonate formations.

Keywords Well logs · Carbonate formation · Facies · Fractures · Vugs · Porosity · Machine learning

1 Introduction and Scope

Wireline logs provide a continuous depth profile of subsurface rock properties, each with a different vertical resolution and depth of investigation. They are obtained by lowering a string of wireline tools into a drilled borehole. Each tool measures a

A. Haagsma · E. Howat · J. Schuetter · A. Conner · B. Grove · S. Mishra (✉) · N. Gupta
Battelle Memorial Institute, Columbus, OH, USA
e-mail: mishras@battelle.org

I. Fukai
University of Tennessee, Knoxville, TN, USA

J. C. de Dios et al. (eds.), *CO_2 Injection in the Network of Carbonate Fractures*, Petroleum Engineering, https://doi.org/10.1007/978-3-030-62986-1_2

unique property of the rock formations into which the boreholes are drilled. Some of the commonly used wireline tools include gamma ray, neutron, density, resistivity, photoelectric effect, and sonic logging tools. Log data from these tools are analyzed to identify formation top depths, characterize the structural and stratigraphic framework of the subsurface, and to calculate basic formation properties such as reservoir thickness, porosity, and fluid saturations. Additionally, there are other advanced wireline logs that provide valuable subsurface data but are less commonly acquired due to higher costs and borehole-specific limitations. Of these advanced logs, this section will focus on pulsed neutron capture, nuclear magnetic resonance, formation micro-imagers, and elemental spectroscopy logs. Each of these logs, both basic and advanced, can exhibit specific responses that may be used to identify and characterize fractured carbonates (e.g. [28]).

This chapter will begin with an overview of wireline logs, followed by a discussion of the types of analyses that can be performed with the information obtained from the logs. This will be followed by a discussion of best practices regrading data QA/QC and analysis workflow. Next, a case study will be presented to demonstrate well log interpretation for carbonate reservoirs. The final section provides a brief introduction to the emerging technique of machine learning applications for well log analysis including an example.

2 Overview of Wireline Logging Techniques

2.1 Types of Logs

This section provides a brief overview of the principles corresponding to basic and advanced wireline logging techniques.

1. *Gamma Ray (GR)*

The gamma ray log measures the natural radioactive decay of elements such as uranium, potassium, and thorium, that are present in in minerals such as potassium-feldspar, clays, and heavy mineral fractions such as zircon, apatite, and monazite. The relative abundance of these radioactive minerals as typically observed in sandstone, carbonate, and shale formations can be correlated to the gamma ray log to identify reservoir and non-reservoir rocks and caprocks for potential injection/storage applications. The log is measured in American Petroleum Institute (gAPI) units. Gamma ray log data can be acquired in both open boreholes and through well casing, and they are often used to provide a complete record of the borehole from the total depth of the well up to the ground surface.

Carbonate rocks consist predominantly of calcite and dolomite with low amounts of radioactive elements. Consequently, carbonate reservoirs generally exhibit a low gamma ray log response (e.g. $\leq$75 gAPI) (Fig. 1). There are some instances, however,

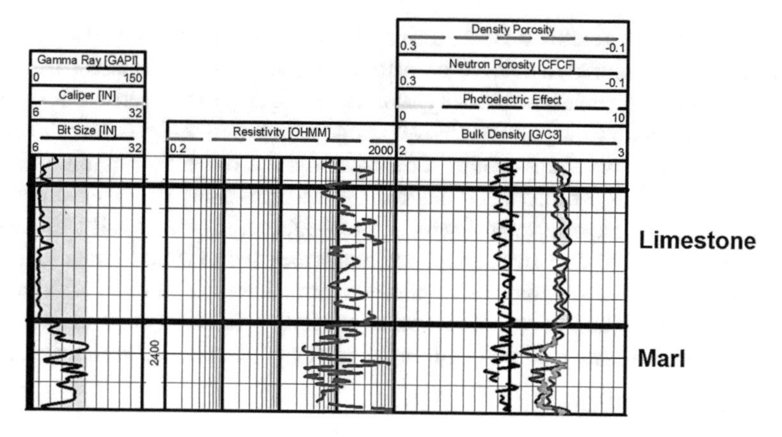

Fig. 1 Gamma ray, resistivity, neutron, density, and photo-electric log curves showing log responses in clean limestone versus marl (limestone-rich shale/mudrock)

wherein carbonate rocks may contain high amounts of radioactive uranium associated with both syngenetic and epigenetic deposition of phosphate minerals and heavy mineral detritus in oxidized or mildly reduced environments [3]. Sometimes in these environments, sudden sharp increases in the gamma ray log may indicate the presence of uranium-rich fracture filling (e.g. hydrothermally-derived fluorite, apatite, phosphorite filling), as uranium tend to strongly partition in the fluid phase in the presence of carbonate ions.

Since different rock types typically have different gamma ray signatures, the gamma ray log can also be used to identify lithologic intervals that may be more prone to fracturing. For example, a shale interval within a clean carbonate reservoir would exhibit a distinctly higher gamma ray signature (e.g. $\geq$75 gAPI) due to radioactive elements in clay minerals, and this shale interval could be targeted to further assess the presence and occurrence of fractures.

2. *Neutron Porosity (NPHI)*

Neutron porosity logs are derived by a neutron source within the wireline tool that measures the effect of the borehole and surrounding formation on the neutron cloud emitted by the tool using two detectors; one placed above the source and one at the bottom of the tool. Collisions between the emitted neutrons and other nuclei within formation minerals/fluids results in neutron energy loss and scattering. Since hydrogen nuclei have approximately the same mass as a neutron, collisions with hydrogen has the greatest impact on neutron scattering and the neutron log can be used to estimate the hydrogen index of the formation. As hydrogen atoms are always present in fluids (e.g., brine, oil) residing in the pores of the reservoir but rarely present in the parent rock materials, the measurement of hydrogen allows estimation of the amount of fluid-filled porosity However, the presence of gas can cause the neutron porosity to decrease, otherwise known as the gas effect. The neutron porosity measurement is calibrated to an assumed matrix value, such as limestone or dolomite, and results are presented in units of porosity for the specific matrix chosen.

The dual-detector neutron porosity tool has a slightly lower vertical resolution (than single-detector tools) and a relatively large depth of investigation with respect to distance away from the wellbore. Consequently, the overall averaging effect of these features makes it difficult to characterize fractures using the neutron porosity log alone due to their aperture and distance from wellbore. Short intervals of increased neutron porosity/porosity spikes could potentially be used to identify open fractures filled with water if corroborated by signatures from other tools. If filled with gas, the neutron could show an interval of decreased porosity spikes. Figure 2 illustrates an example of an increase in porosity attributed to vuggy zones filled with brine.

3. *Bulk Density (RHOB)*

The density wireline log tool measures the bulk density of a formation based on the reduction in gamma ray flux between a source and a detector associated with a phenomenon known as Compton scattering (scattering of a photon by a charged particle, which can cause changes in energy). It is calculated from a reference grain density associated with an assumed mineralogy. For carbonates, a grain density of 2.71 g per cubic centimeter (g/cm^3) is applied to limestone formations based on the grain density of calcite, and a density of 2.87 g/cm^3 is applied to dolomitic formations based on the grain density of dolomite.

The difference between the wireline-measured bulk density and assumed grain density can be attributed to void space in the reservoir and used to calculate density porosity. If the density log exhibits large density correction values associated with decreases in bulk density and/or large increases in porosity that are not observed on the neutron log, a fracture may be present. Borehole rugosity/washouts can affect the log data by widening the borehole. Such zones should be identified using the caliper curve as it can produce erroneous data.

4. *Resistivity (RT)*

Resistivity (RT) logs measure the resistance to flow of electrical current in subsurface formations and fluids. Since water readily conducts electrical current, low resistivity can be indicative of a large volume of interconnect brine-filled pores within a carbonate reservoir. Resistivity response can be used to characterize fluid type and saturations in matrix pores and fractures. Resistivity logs can be acquired at three depths of investigation associated with increasing depth away from the borehole: shallow, medium and deep. Since the drilling mud within the borehole is typically higher resistivity than the resistivity of in-situ formation brines, the shallow resistivity curve generally exhibits higher values relative to the deeper curve.

Cross-over between the shallow and deep resistivity curves, in which the shallow resistivity log measurement is less than the deep resistivity log, can be used to identify the presence of fractures due to mud invasion within the fractures. Micro resistivity logs, also referred to a micrologs, may be used to indicate fractures by showing low resistivity spikes opposite open fractures, and high resistivity spikes for healed/cemented fractures.

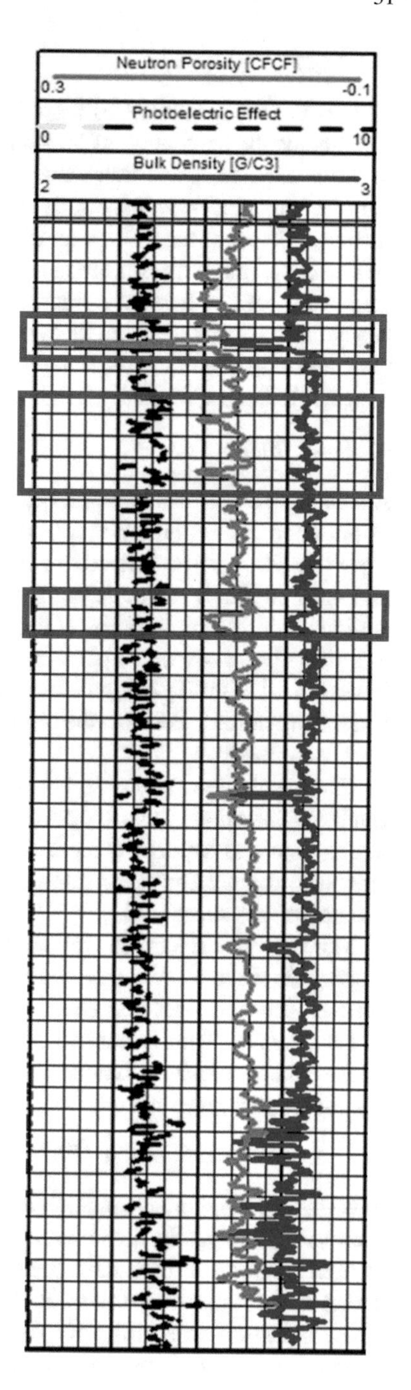

Fig. 2 Example neutron porosity log (green curve) from a carbonate reservoir in the Appalachian basin showing sudden, positive increases in porosity over known vuggy zones

5. *Photoelectric (PE)*

The photoelectric index (PE) log is recorded as a part of the density measurement and records the absorption of low-energy gamma ray in barns/electron. A PE value of 5.0 is expected for a limestone and 3.1 for dolomite. Large PE values, greater than 5.0 barns/electron, especially when weighted muds are used, can be used as a fracture indicator. Barite, for example, \ has a very large photoelectric cross section, 267 as compared with limestone or dolomite. Thus, the PE curve should exhibit a very sharp peak in front of a fracture filled with barite loaded mud cake. However, heavy muds can also plug up the well. Corroboration from other data is essential for more accurate identification. In light weight muds, an abnormally low PE value, less than 1.7, indicates, fractures, bad hole condition, or coal.

6. *Sonic (SON)*

A sonic tool emits an acoustical signal into the rocks and measures the travel speed of the wave echoes through the formations. Both compressional (P) and shear (S) wave velocities can be derived from sonic logs to calculate geomechanical properties and stress regimes of the subsurface. Sonic logs can also be used to provide an additional derivation of porosity for reservoir characterization. Typical travel times in a limestone is 49 or 44 μs/ft in a dolomite. Drilling muds have a much higher travel time, 180–190 μs/ft, which can increase the overall arrival time of a carbonate if it fills fractures. Likewise, the presence of hydrocarbons can also cause the arrival time to increase.

Historically, the difference between the acoustic porosity and neutron porosity can be attributed to secondary porosity influence such as vugs and fractures. This is due to the acoustic porosity representing the matrix porosity while neutron porosity represents the bulk formation total porosity. The results of this application are variable and not always successful.

More advanced techniques are available with sonic (acoustic) logs which can help identify fractured and/or vuggy zones. This includes the evaluation of anisotropy between the slow and fast shear wave arrivals and deep shear wave processing. Anisotropy can identify anomalies which cause differences in shear arrival times. These features can be attributed to fractures and vugs. Deep shear wave processing can identify fractures near the wellbore and up to 200 ft away (Fig. 3).

7. *Pulsed Neutron Capture (PNC)*

Pulse neutron logging technology is used to identify the presence of reservoir fluids such as oil and gas in the borehole [1]. This tool operates by generating a pulse of high energy neutrons, subsequently measuring the neutron decay over time and across a wide energy spectrum. PNC can be ran in both open and cased boreholes. The tool measures the ability of an element to capture thermal neutrons and generates a log of this value, known as the thermal neutron capture cross section, or Sigma [5]. Thermal neutron capture measurements are compared to referenced values of common downhole fluids and formation matrices. A higher sigma value equates to a greater ability of an element to capture, or absorb, the neutrons. Formation brines and

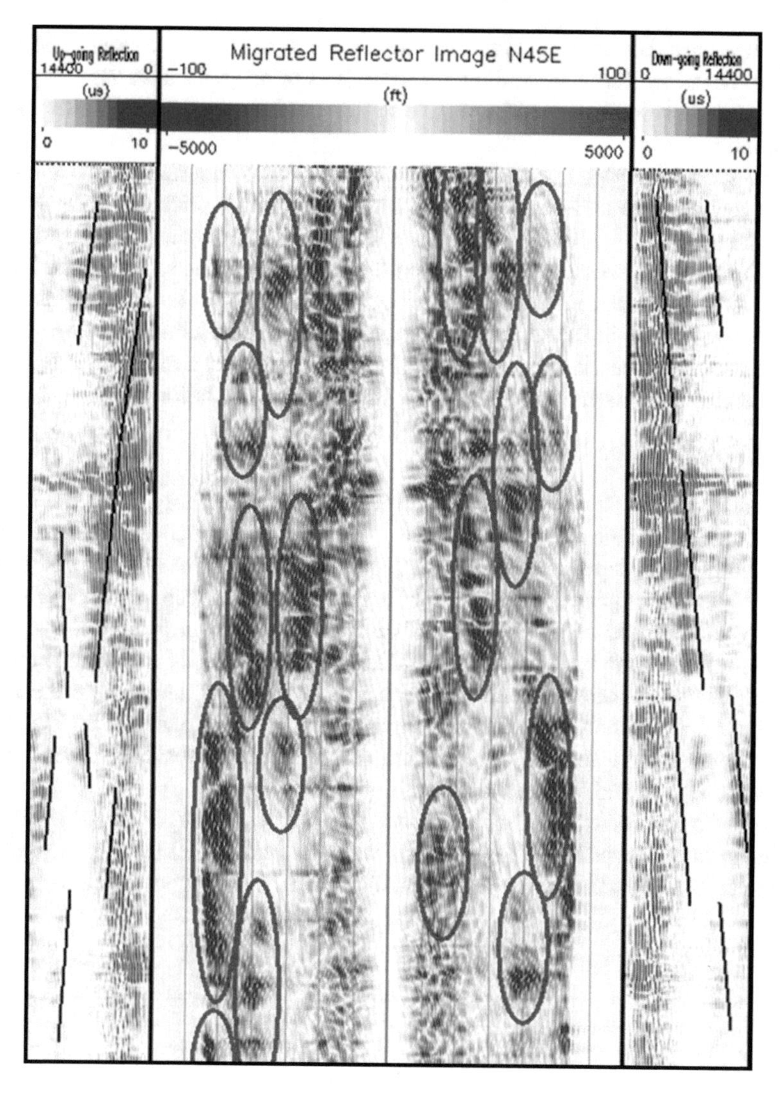

Fig. 3 Example of deep shear wave processing showing an identified fractured zone

oil all have distinctive sigma values which can be used to determine fluid saturations at various depths surrounding the borehole. However, the tool does not have the resolution to distinguish between multiple low-density compounds/elements such as methane (CH_4) and carbon dioxide (CO_2) in supercritical or gas phase due to similar sigma values.

Measured data from pulsed neutron logging is processed and analyzed using a technique known as sigma analysis. This is based on a material balance equation which defines the measured formation sigma as the weighted sum of the component sigma responses for fluids and lithology within the measurement volume. The equation can be solved for water and hydrocarbon (gas or oil) saturation using inputs of porosity, shale volume and representative sigma values for the specific fluids present in the formation.

Well-specific modeling coupled with vendor-specific proprietary analysis techniques can analyze and compute multi-phase saturations, including oil, gas and water saturation in wells. For example, the graphical representation described in Fig. 4, utilizes the measured data in Monte Carlo simulations to determine formation fluid components. It also establishes quantitative interpretations for fluid saturations percentages in the formation. Furthermore, bore-hole and porosities conditions can create uncertainties, however utilization of advanced modeling can reduce uncertainties and provide better interpretations of fluid saturations [11].

8. *Nuclear Magnetic Resonance (NMR)*

Nuclear Magnetic Resonance (NMR) well-logging technology is used for pore system characterization and is sensitive to the fluids contained within rock while being lithology independent. NMR logging offers a measure of how porosity is divided between larger pores where fluid is free to move (effective porosity) versus porosity attributed to clay-bound water which is essentially trapped, often referred to as the bulk volume irreducible (BVI). Thus, effective porosity is the total porosity minus BVI. NMR measurements are frequently used for producing a permeability log, either through the Coates equation or the Kenyon (SDR) permeability equation. NMR measurements also play a role in the determination of residual oil saturation, capillary pressure and in facies analysis.

Most of the NMR theory, practice, and case studies are reported for reservoirs dealing with matrix porosity. Lesser work has been done for NMR logging applications in fractured and/or vuggy reservoirs. Xiao and Li [34] conducted a theoretical

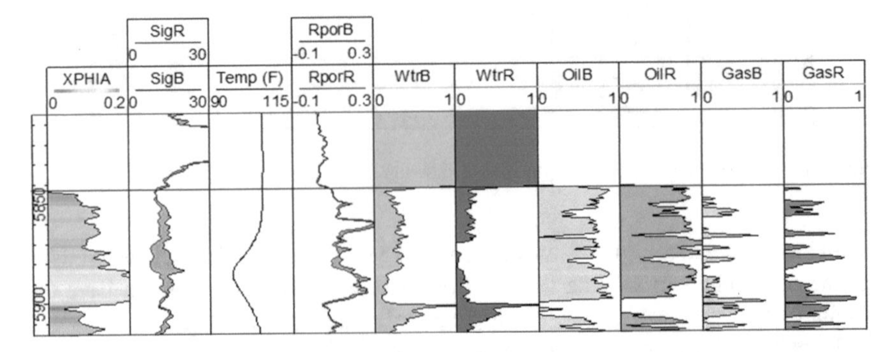

Fig. 4 Baseline (B) and repeat (R), logged saturations for the well. Column 1 shows a neutron porosity and bulk density derived porosity log. SigB (baseline) and SigR (repeat) in Column 2 is the sigma measurement. Column 3 is the temperature log measurement logged after the repeat pulsed neutron measurements. Column 4, RporB and RporR are the baseline and repeat PNC porosity measurements. Columns 5 and 6 show initial and repeat water curves. Columns 7 and 8 show initial and repeat oil. Columns 9 and 10 show initial and repeat gas data

characterization of NMR logging response in fractured reservoirs. Their findings show that NMR logging is sensitive to fracture characteristic such as aperture, dip angle, and intensity. It is also sensitive to drilling fluids and antenna length. In actual practice they advise that NMR logging in fractured reservoirs would require the integration of other logging information such as fullbore formation micro imaging (FMI). Altogether, the NMR response would be affected by the fractures, the matrix, the drilling fluids, and on whether the formation was oil-bearing or wet.

To make the NMR more valuable for a carbonate reservoir evaluation, it is recommended to calibrate with core measured properties such as porosity and permeability and using pore size distributions to determine appropriate cutoffs. In other cases, core measurements can be made in a laboratory to further calibrate the NMR log.

9. *Image Log*

Image logs provide a visual representation of the wellbore environment. These tools use either a rotating sensor to measure acoustic impedance images around the circumference of the borehole or a pad tool with electrodes to produce a resistivity image of the formation within the borehole that can be used to qualitatively and quantitatively assess fractures, bedding planes, paleocurrent features, vugs and other pore types on a very fine scale (Fig. 5). Fractures tend to produce a high contrast anomaly due to fluid invasion and wellbore breakout. Vugs can produce a mottled appearance with

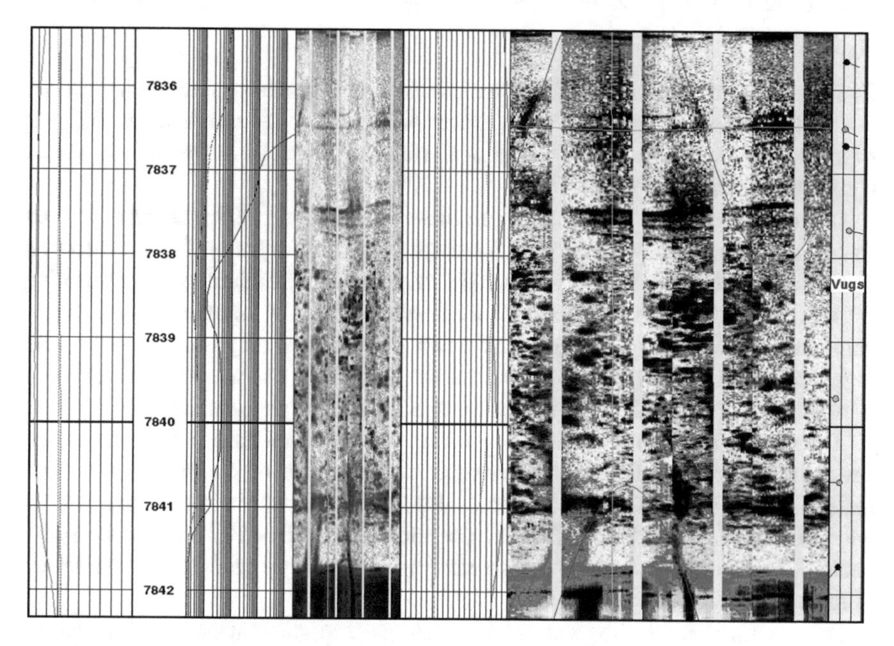

Fig. 5 Example image log over a carbonate interval showing highly vugular zone and fracture near the base

high contrast to matrix rock. When image logs are run with an orientation tool, fracture analysis can be enhanced by analyzing the dip and orientation of the fractures, which are critical characteristics to understand when modeling fractured reservoirs.

10. *Elemental Spectroscopy*

Elemental spectroscopy logs are used to analyze in-situ geochemical and mineralogical composition of subsurface rock formations and can provide matrix (grain) density curves for more accurate porosity estimation. One type of elemental spectroscopy log is the natural gamma ray spectroscopy log. This log measures the individual emissions of gamma rays from radioactive decay of thorium, uranium, and potassium, providing three curves representing the relative abundance of each element in the formation. Similar to the more common gamma ray log previously described, an increase in the uranium concentration from the natural gamma ray spectroscopy log can indicate the presence of fractures.

2.2 Types of Analyses

This section describes the types of analyses that can be performed with the various logging techniques described above for characterizing the formation of interest.

1. *Facies*

Facies are discrete intervals within a rock formation which have similar characteristics such as lithology, porosity, and structures. These can be used to better interpret a reservoir and its property distributions. Certain facies can be assigned to represent vuggy and/or fractured zones.

One common method to divide a carbonate into facies is distinguishing between shaley intervals and clean carbonate. A maximum gamma ray cut-off of 75 American Petroleum Institute gamma ray units (gAPI) is an industry standard for distinguishing between clean sandstone and carbonate reservoirs with low amounts of radioactive components (GR <75 gAPI), and non-reservoir shale and clay-bearing rocks with higher proportions of radioactive constituents (e.g., [30]). At a minimum, this should be applied to distinguish between clean and shaley/clay-rich carbonates.

The determination and inclusion of facies can be beneficial for formation evaluation and modeling. However, carbonate facies tend to be composed of similar material (especially on log resolution scale) and can be difficult to distinguish on wireline logs. The correlation of whole core and/or sidewall cores in conjunction with geologic knowledge of the depositional environment can help guide the determination of facies.

It is common for carbonates to have undergone significant diagenesis, such as dolomitization and dissolution. Where facies are good representatives of the original depositional environment, the diagenetic footprint could render these unimportant for formation evaluation and modeling purposes. Careful consideration should be given to the influence of diagenesis on a carbonate reservoir and the development of representative facies.

Alternatively, the use of electrofacies can be used. These are statistically derived intervals of like-log signatures. The electrofacies can be predicted using available data and linked to relevant geologic features. Where depositional facies could have similar log distributions, electrofacies will be unique. Figure 6 illustrates derived electrofacies and lithology-based facies. The electrofacies further divide the reservoir and identifies higher porosity units, while the lithology-based facies do not capture that variability.

2. *Shale Volume*

When evaluating a carbonate formation, it is important to estimate the volume of shale to better define the reservoir. Shale can overestimate the porosity of a system, leading to inaccurate volumetric calculations. The fraction of shale in a given formation can be estimated using the normalized GR curved via Eq. (1):

$$V_{shale} = \left(GR_{log} - GR_{res}\right) / \left(GR_{shale} - GR_{res}\right) \tag{1}$$

where

V_{shale} = shale fraction (i.e. non-reservoir)
GR_{log} = gamma ray log value
GR_{res} = gamma ray value of clean sandstone and/or carbonate in each zone
GR_{shale} = gamma ray value from a nearby shale interval
$1 - V_{shale}$ = clean sandstone and/or carbonate fraction (i.e. reservoir).

3. *Porosity*

Total porosity is defined as the combined percentage of interconnected, isolated, and clay-bound porosity of the total formation. Effective porosity is defined as the combined percentage of interconnected and isolated porosity in sandstone and/or carbonate reservoirs and does not include the clay-bound porosity associated with the non-reservoir, shale fraction. Total and effective porosity can be calculated from neutron, bulk density, and sonic log data, both from individual logs and a combination of multiple logs. Density porosity can be calculated from Eq. (2):

$$\phi_D = (\rho_{ma} - \rho_b) / (\rho_{ma} - \rho_{fl}) \tag{2}$$

where

ϕ_D = density porosity
ρ_{ma} = matrix density
ρ_b = bulk density (from the density log)
ρ_{fl} = fluid density.

If direct matrix density measurements are not available for the formations of interest (whole and/or sidewall core analyses), empirically derived industry standards

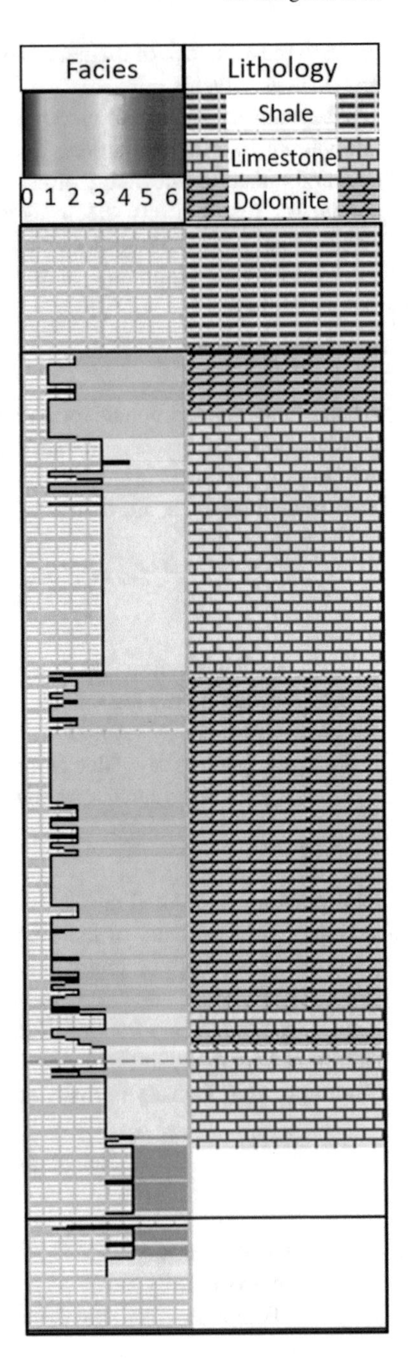

Fig. 6 Example well showing electrofacies (left track) and rock facies (right track) from the Michigan Basin

of 2.65 g/cm^3, 2.71 g/cm^3, and 2.87 g/cm^3 can be assumed for sandstone, limestone, and dolomite formations, respectively [2].

The neutron logging tool is commonly used and produces a log (NPHI) of total porosity (ϕ_{tot}). Total porosity is the portion of a rock that is occupied by fluids such as oil or water. Total porosity is comprised of a free fluid volume (FFV), capillary-bound water, and clay-bound water. The free fluid volume is considered effective porosity (ϕ_e) because this represents the pore space that is accessible, either through the production of oil or for the storage of fluids such as wastewater or supercritical CO_2.

Sonic porosity can be calculated using Eq. (3):

$$SPHI = (\Delta T_{log} - \Delta T_{matrix})/(\Delta T_{fluid} - \Delta T_{matrix}) \quad (3)$$

where

$SPHI$ = acoustic (sonic) porosity
ΔT_{log} = acoustic travel time log value
ΔT_{matrix} = acoustic travel time of rock matrix
ΔT_{fluid} = acoustic travel time of pore fluids.

If direct measurements are not available for equation parameters, a standard acoustic travel time of 47.6 microseconds per foot (μs/ft) can be assumed for ΔT_{matrix} in calculations for limestone formations, and a ΔT_{matrix} value of 52.6 μs/ft can be assumed for consolidated sandstone formations [7]. Formations containing predominately brine can be assumed to have an ΔT_{fluid} value of 189 μs/ft [7].

The estimated reservoir fraction ($1 - V_{shale}$) can be used to correct the total porosity curve for shale/clay effects and generate effective porosity logs. Effective porosity was estimated by calculating the porosity associated with the reservoir fraction via Eq. (4):

$$PHI_e = PHI_t \times (1 - V_{shale}) \quad (4)$$

where

PHI_e = effective porosity
PHI_t = total porosity
$1 - V_{shale}$ = sandstone and/or carbonate fraction (i.e. reservoir fraction).

4. *Permeability*

Permeability can be estimated from wireline logs such as resistivity, and nuclear magnetic resonance, when core data is not available, but is an indirect measurement. Permeability estimated from resistivity logs have shown to be successful in sandstone formations [18], but are less successful in carbonates due to complex pore geometry and secondary porosity features. NMR is a better alternative for estimating permeability in carbonate formations. This is best done in conjunction with core analysis to calibrate the log measurements, otherwise the results can be magnitudes off [32].

Permeability from core data can also be plotted as a function of core porosity data to determine the relationship between the two parameters, and the resulting regression can be used to derive transform equations to estimate permeability from porosity log data.

Secondary porosity features such as vugs and fractures are important reservoir features and can enhance the overall permeability of the formation. While permeability can be derived from porosity for matrix porosity, permeability in secondary porosity is typically unpredictable. The presence of fractures and vugs can cause high and oftentimes immeasurable permeability. While permeability can be derived from porosity for matrix porosity, permeability in secondary porosity is typically unpredictable. In these cases, it is more important to understand the matrix permeability as it will control the total flow of the reservoir system and be enhanced by the secondary porosity.

5. *Fluid Saturation*

In-situ formation fluid types and saturations are commonly estimated from resistivity and nuclear magnetic resonance logs. Additionally, pulsed neutron capture logs can be used to estimate saturations of oil, gas, and water. Since logging methods are indirect measurements, it is recommended to corroborate the results with core measured values.

Traditionally, water saturation calculations can be performed using the Archie equation, a standard oil and gas formula shown in Eq. (5).

$$S_w = \left(\frac{a * R_w}{R_t * \emptyset^m} \right)^{\frac{1}{n}} \tag{5}$$

where

S_w = water saturation of the uninvaded zone (%)
R_w = formation water resistivity (ohm-ft)
R_t = formation resistivity (ohm-ft)
$\emptyset$ = porosity (%)
a = tortuosity factor
m = cementation exponent
n = saturation exponent.

Values for formation resistivity (R_t) and porosity (ø) are derived from the wireline logs. Formation water resistivity (R_w) value can be determined based on salinity and temperature using an industry chart, or fluid measurements. The constants a, m, and n can vary greatly in carbonates depending on cementation and pore structures. The presence of vugs and fractures can also impact the values. A common industry standard for carbonates is a = 1, m = 2, and n = 2. Laboratory measurements can also be conducted to determine reservoir and/or facies specific variable values for these coefficients.

Alternatively, the a, m, and n values can be treated as variable. This can be done using facies or electrofacies to develop specific variables for individual zones. For example, electrofacies A could have different a, m, and n values than electrofacies

B. Using a variable approach would better represent the variability of the reservoir than using an average approach.

6. *Crossplot Analysis*

Crossplot analyses are valuable, quick look approaches to identify zones of interest, determine lithology, develop reservoir flags, and estimate crossplot porosity. A common method is to plot the neutron porosity versus the bulk density (Fig. 7). Common features observed in carbonates are outlined in Fig. 7 (albeit without the samples to clearly show the different zones). Dolomite is outlined in purple and limestone in blue. The green dashed polygon represents areas of crossplot porosity greater than 5%. This can be modified for the interpreters determined porosity threshold of interest. Additionally, salt trends are highlighted in red and anhydrite in orange. Other useful plots include density-sonic, gamma ray-density, resistivity-porosity, and geomechanical properties.

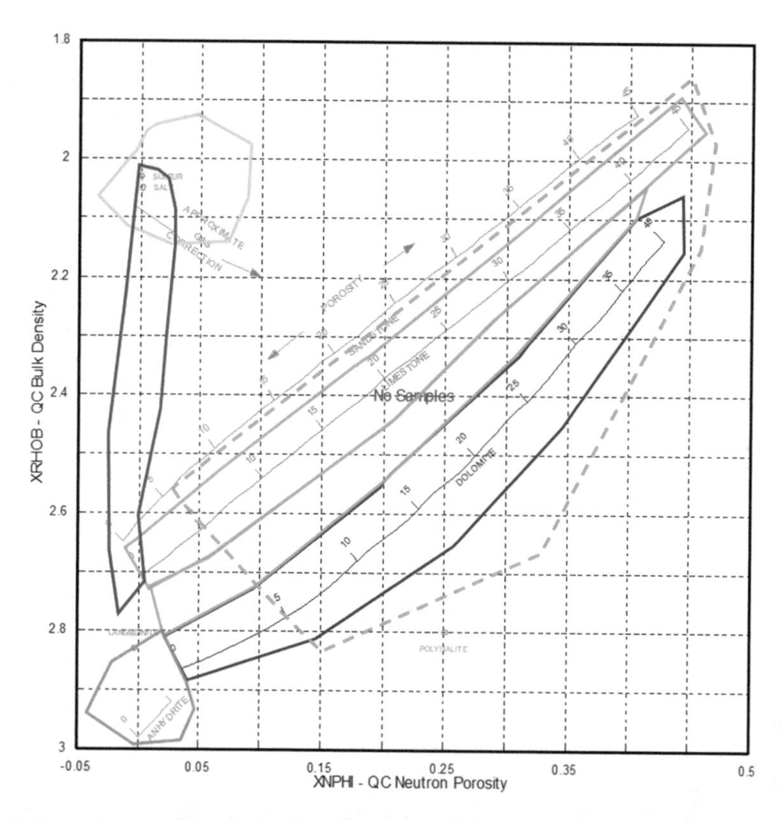

Fig. 7 Example crossplot analysis method using neutron porosity and bulk density to identify lithology and porosity zones of interest (see Fig. 11 for an example with actual samples)

7. *Geomechanics*

Geomechanical properties are important to characterize a reservoir for safe injection practices but can also be valuable when evaluating carbonate formations for fractures. It can be assessed using a combination of wireline logs including bulk density and dipole sonic to compute geomechanical properties including bulk modulus, shear modulus, Young's modulus, and Poisson's ratio. Additionally, image logs can be used to derive geomechanical parameters of fractures and borehole breakout.

Dynamic shear modulus evaluates a formation's rigidity or resistance to shear stress. It can be calculated used bulk density and shear wave slowness as shown in Eq. (6).

$$G = A\frac{\rho}{DT_s^2} \tag{6}$$

where

$G =$ dynamic shear modulus (10^6 psi)
$A = 1.3476 \times 10^4$ (conversion parameter, with units of 10^6 psi/((g/cc)/(μs/ft)^2)
$\rho =$ bulk density (g/cc)
$DT_s =$ shear wave slowness (μs/ft).

The bulk modulus of a rock formation expresses the resistance of the formation to compression. This is measured using compressional-wave and shear-wave slowness, as expressed in Eq. (7).

$$K = \left(1.3476 \times 10^4\right)\rho\left(\frac{1}{DT_C^2} - \frac{4}{DT_s^2}\right) \tag{7}$$

where

$K =$ bulk modulus (10^6 psi)
$\rho =$ bulk density (g/cc)
$DT_C =$ compressional wave slowness (μs/ft)
$DT_s =$ shear wave slowness (μs/ft).

Poisson's ratio indicates the relationship between lateral and axial strain of the rock formation, as expressed in Eq. (8).

$$V = \frac{1\left(\frac{DT_s^2}{DT_C^2}\right) - 2}{2\left(\frac{DT_s^2}{DT_C^2}\right) - 1} \tag{8}$$

where

$V =$ Poisson's ratio (dimensionless)
$DT_c =$ Compressional Wave slowness (μs/ft)
$DT_s =$ Shear Wave slowness (μs/ft).

And finally, the Young's modulus is a measure of the stiffness of a material, or the resistance to elastic deformation, and is calculated using formation density, shear-wave velocity, and Poisson's ratio, as is expressed in Eq. (9).

$$E = 2G(1 + V) \tag{9}$$

where
E = Young's Modulus (psi)
G = dynamic shear modulus (psi)
V = Poisson's ratio (dimensionless).

8. *Fractures and Vugs*

The identification and characterization of vugs and fractures require the use of advanced wireline logs and/or whole core. Image logs are often used to visually identify zones which are fractured and vuggy. Orientation and dip information can also be derived for fractures from image logs when they are run in conjunction with an orientation tool. Deep shear wave processing can also be used to identify significant fractures both near and far from the wellbore, up to 200 ft away.

Where advanced logs and whole core are not available, machine learning techniques can be used to develop predictive models of where the secondary porosity features are occurring. Section 5 provides an overview of the approach and an example application.

3 Best Practices

3.1 Data Quality Assurance/Quality Control (QA/QC)

1. *General Considerations (Depth Correlation)*

To confidently use and correlate whole core data, it is best to depth correct it to match the wireline logs. Whole core can be as much as 30 ft different in depth compared to logs, which is why corrections are essential. There are different methods that can be done including: (1) depth comparison of a distinct marker or formation such as shale bed or change in lithology, (2) visually shifting of data to match core to wireline log characteristics (porosity, gamma ray, bulk density), or (3) statistical matching of properties to optimize correlation. Once depth is corrected, the whole core data can be correlated with wireline logs.

Figure 8 shows an example well from the Michigan Basin. The original depth of the core measured porosity had a distinct spike in porosity at 4065 ft, which was seven feet from the same distinct spike in log measured porosity. By applying the seven-foot shift, the corrected core porosity then matched the log porosity.

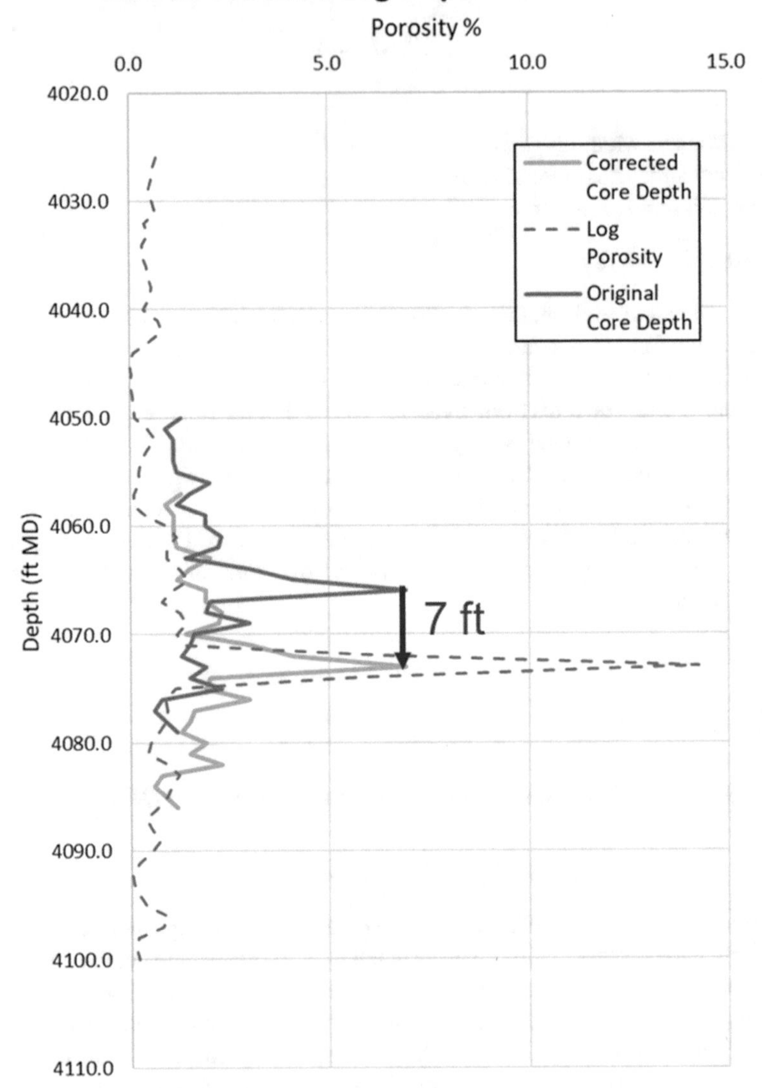

Fig. 8 Example of depth correction between core porosity (orange) and log porosity (blue) to improve correlation between data types

2. *GR Normalization*

As previously mentioned, a maximum gamma ray cut-off of 75 gAPI is an industry standard for distinguishing between carbonate reservoirs with low amounts of radioactive components (gamma ray values <75 gAPI), and non-reservoir shale and clay-bearing rocks with higher proportions of radioactive constituents (e.g., [30]). In order to use gamma ray log data for facies analysis and shale volume calculations, it's recommended to first normalize the data in order to eliminate varying signal intensities and attempt to establish consistent readings for carbonate versus feldspathic sandstone and shale lithologies.

Gamma ray logs can be normalized based on statistical data from one of more selected type-wells in the study area. The type-wells selected should exhibit GR log signatures that generally fall within the normal range observed in sedimentary rocks (~10 to 200 gAPI), with values greater than 75 gAPI consistently observed in formations known to consist of shale and mudstone and values less than 75 gAPI observed in known sandstone and carbonate intervals. The mean GR values from non type-wells can be normalized to the mean calculated for each formation or zone interest in the type well(s) via Eq. (10):

$$GR_{norm} = GR_{log} - GR_{zonemean} + GR_{normmean} \tag{10}$$

where

GR_{norm} = the normalized gamma ray log

GR_{log} = the original gamma ray log value

$GR_{zonemean}$ = the mean gamma ray value for the zone of interest in the non-type well

$GR_{normmean}$ = the mean gamma ray value for the zone of interest in the type well.

The ranges of the normalized GR curves can then be calculated for each formation/zone of interest to identify outliers in the dataset requiring additional normalization using the minimum and maximum GR values observed in type wells. Normalization by minimum and maximum values can be conducted via Eq. (11):

$$GR_{norm} = GR_{normmin} + \left(GR_{log} - GR_{zonemin}\right) \times (GR_{normmax} - GR_{normmin})/(GR_{zonemax} - GR_{zonemin}) \tag{11}$$

where

GR_{norm} = the normalized gamma ray log

$GR_{normmin}$ = the minimum gamma ray value for the zone of interest in the type well

GR_{log} = the original gamma ray log value

$GR_{zonemin}$ = the minimum gamma ray value for the zone of interest in the non-type well

$GR_{zonemax}$ = the maximum gamma ray value for the zone of interest in the non-type well.

This normalization procedure will help to ensure mineralogically homogenous (i.e., clean) carbonates have gamma ray values less than 75 gAPI and shale/clay-rich intervals have values above 75 gAPI. Figure 9 illustrates an example from a well in the Michigan Basin. The left image is pre-corrected and pre-normalized gamma ray and the right image is post-corrected and normalized gamma ray log distinguishing between the Utica Shale and Trenton carbonate.

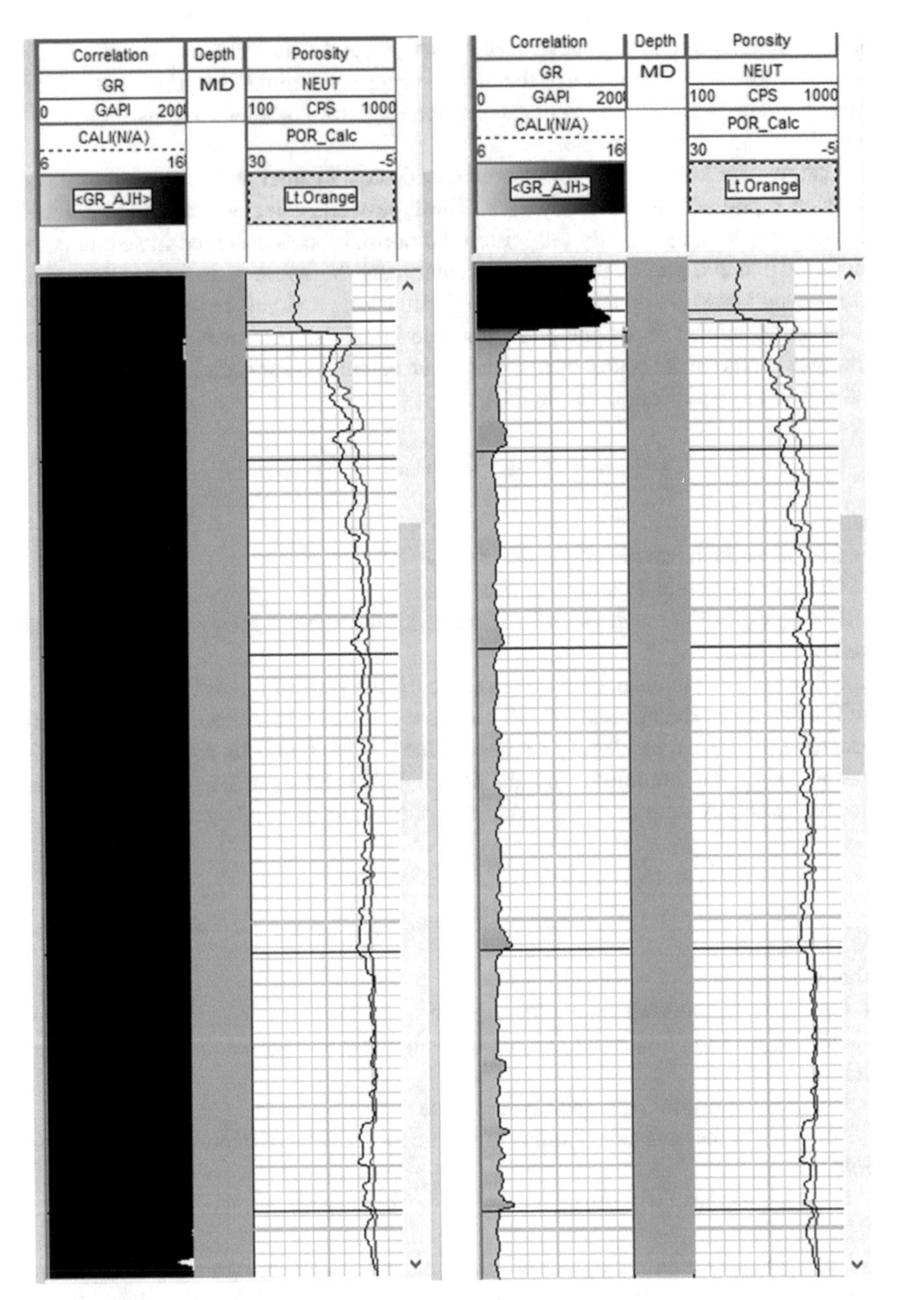

Fig. 9 Example results of a corrected and normalized gamma ray log from a well in the Michigan Basin showing a distinction between the Utica shale and Trenton carbonate (right image) which was previously indistinguishable (left image)

3. *NMR Calibration*

For a quantitative estimate of effective porosity, an accurate placement of the T2 cut-off is necessary for calibrating the NMR log. This cutoff can vary based on lithology and diagenesis of the pore system. For example, different $T_{2\text{cutoff}}$ values are required when logging in carbonates because surface relaxivity in carbonates is weaker than in sandstones, resulting in slower relaxation rates (longer T_2). Accurate placement of the T2 cut-off in fractured carbonates can be determined from core-derived NMR and mercury-injection capillary pressure (MICP) measurements.

For example, in the case study provided above, core plugs were collected from the whole core samples for MICP analysis. The resulting pore throat systems were evaluated to determine dominant systems. These were applied to the Timur-Coate model to refine the permeability estimates [9]. This reduced the originally estimated NMR permeability to be within range of core measured permeabilities (Fig. 10).

4. *Salt and Anhydrite Flagging*

The presence of salt and anhydrite can cause a decrease in reservoir quality because it can plug porosity, and is common in some carbonate reservoirs. Bulk density, PE, and elemental spectroscopy logs are valuable tools to identify and flag intervals that are affected. Salt has a low grain density of 2.04 g/cm^3, where dolomite is on average 2.83 g/cm^3 and limestone is 2.71 g/cm^3. The influence of a lower-density mineral, such as salt, in a carbonate will lower the total bulk density of the rock. Anhydrite has a higher grain density of 3.0 g/cm^3 which will increase the overall bulk density of a system. However, be cautious if relying on bulk density alone as porosity and fluids also cause changes in bulk density that can appear as salt. A simple crossplot analysis is a quick tool to flag the presence of salt and anhydrite (Fig. 11). The additional of PE and elemental spectroscopy would validate the interpretation as each has unique values for evaporites.

3.2 Analysis Workflow

A general petrophysical analysis workflow typically integrates all available core and log data to characterize key subsurface properties for characterization of storage reservoirs, such as porosity, permeability, and lithofacies as shown in Fig. 12.

4 Well-Log Interpretation Case Study

This section describes a case study of wireline log interpretation in a complex carbonate reef play in the northern Michigan Basin. Multiple CO_2-EOR reefs were studied under the Midwest Regional Carbon Sequestration Partnership (MRCSP)

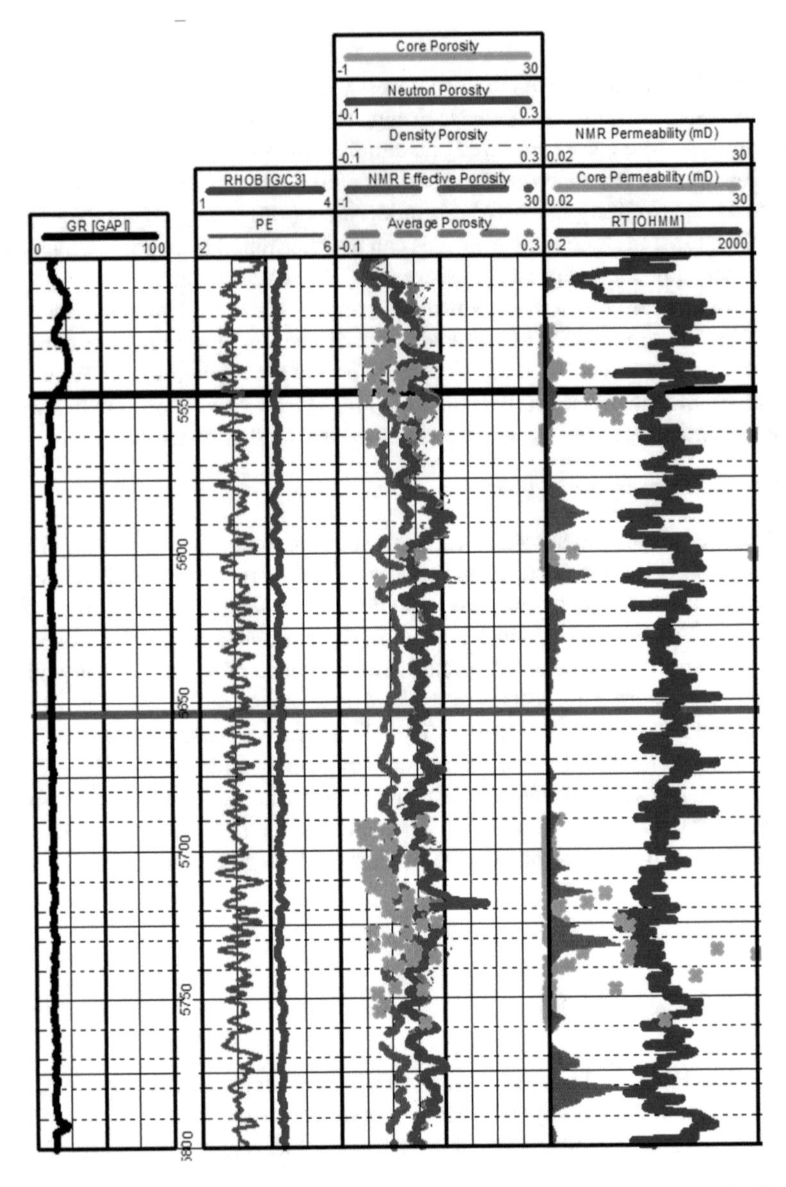

Fig. 10 Example results of calibrating NMR measured permeability using pore throat distributions from MICP data. The right most track shows the NMR permeability in red, after calibration, and the core measured permeability in orange

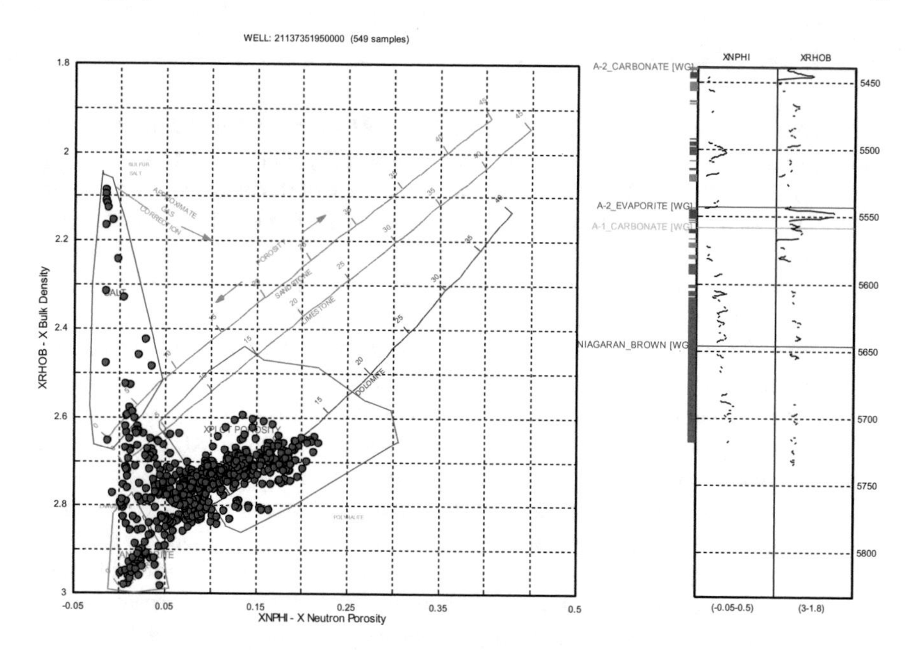

Fig. 11 Example cross plot analysis flagging salt in blue and anhydrite in green

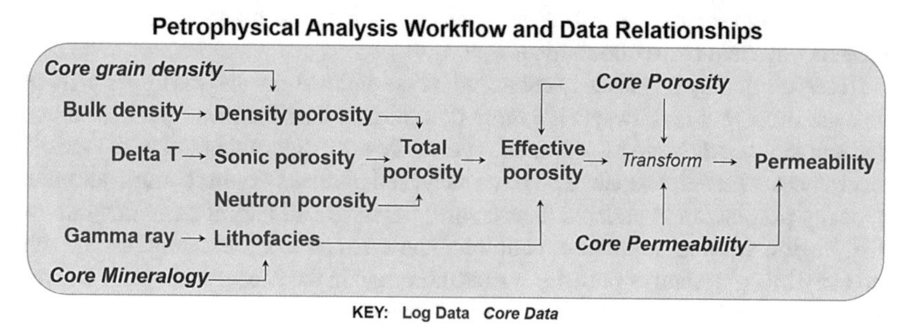

Fig. 12 Schematic showing example data relationships and integration approach for calculating petrophysical properties of potential storage reservoirs

project to demonstrate the safe injection and storage of CO_2 into a carbonate reservoir. The partnership created an opportunity to collect basic and advanced wireline logs, perform reservoir tests, and analyze core data to characterize the reservoir and confining systems.

4.1 Study Area and Geology

Intro sentence needed Upper Silurian carbonate platforms developed along arches that separate the Michigan, Ohio, and Illinois basins [25]; and the Northern Pinnacle Reef Trend (NPRT) developed along the northern slope of the increasingly restricted Michigan Basin [6, 15]. More than 800 individual reefs have been identified in the NPRT. The reefs range from 2000 to over 6000 ft deep, with producing reefs mostly occurring at depths of 3500–5000 ft. Individual reefs are closely spaced and compartmentalized from the enclosing rock; and average 50–400 acres in area, up to 700 ft in height, with steep flanks of 30°–45°, thus fitting the definition of pinnacles by Shouldice [29]. Larger reefs commonly consist of coalesced individual pinnacles and mounds, and height of reefs generally increases basinward.

The resulting carbonate buildups may be completely dolomitized, essentially all limestone, or a heterogeneous mix. Lithology trends with the basin with increasing limestone content towards the basin and increasing dolostone towards the shelf. Porosity types include primary framework voids and interparticle porosity systems, and secondary vuggy, cavernous and fracture porosity as well as intercrystalline and microcrystalline porosity [26]. Porosity values of NPRT reefs average about 3–12%, with the best porosity and permeability associated with dolomitized reef core and flank facies [14]. The most productive reservoir rocks are characterized by well-developed intercrystalline and vuggy porosity with average permeability values of 3–10 millidarcies (mD). Permeability can be significantly higher where fractures intersect vugs and matrix porosity.

Reservoir quality is generally enhanced by dolomitization, and the upper parts of reefs are often, but not always, are more dolomitized than the lower parts. Average reef porosity trends higher towards the shelf following dolomitization with isolated occurrences where limestone reefs have increased porosity. Hydrothermal dolomite is locally present, and is related to structure, fractures and migration of deep fluids [13]. Regionally, non-hydrothermal dolomitization of reefs increases updip, and salt and anhydrite plugging of porosity is more common in the deeper reefs [12]. Degree of salt and anhydrite plugging can greatly decrease reservoir quality.

As part of the Midwest Regional Carbon Sequestration Partnership (MRCSP), several reefs have been studied for CO_2-EOR (Enhanced Oil Recovery) including detailed geologic characterization, modeling, and monitoring in the northern Niagaran pinnacle reef trend in Michigan. The studied reefs are in different stages of CO_2-EOR including late-stage (post primary and tertiary production), on-going (actively producing), and new EOR reefs. The MBF A reef is a late-stage reef which has undergone primary production and CO_2-EOR production while the MBF B reef is a new EOR reef which has undergone primary production and recently started CO_2 injection (Fig. 13). As part of the characterization efforts, new wells were drilled in the two reefs to collected basic and advanced wireline log data, whole core and side-wall cores, and various well tests. This section describes the analyses and results for two reefs with differing lithology and EOR status.

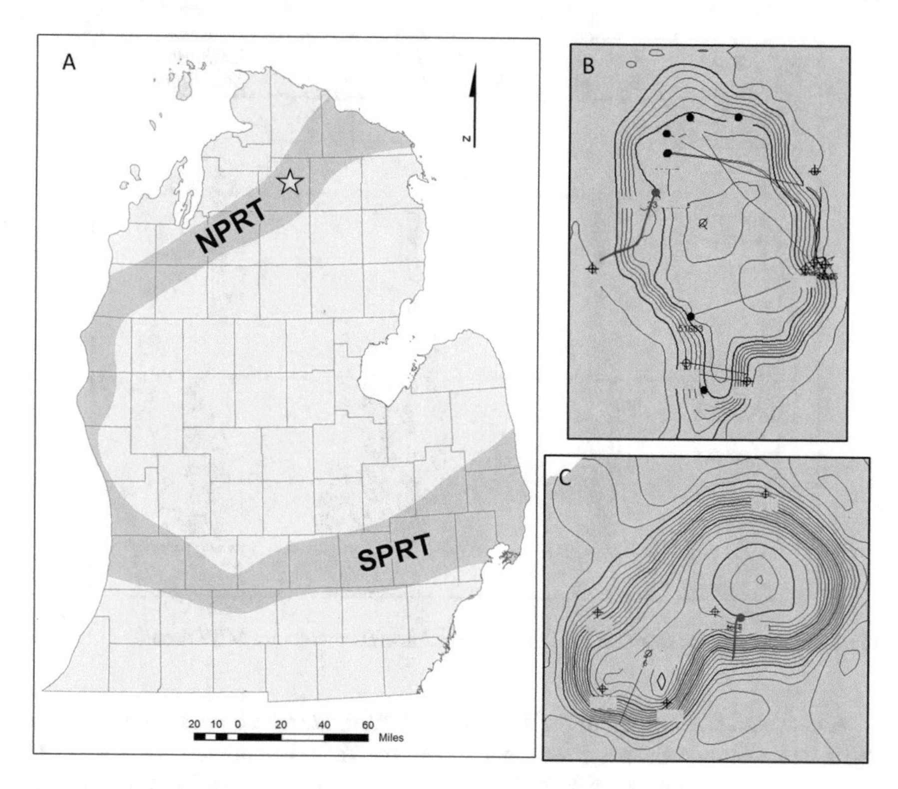

Fig. 13 Map location and structural surfaces of the MBF A and MBF B reefs highlighting study wells in red—both are located in the NPRT within the yellow star

4.2 Description of Data

A new well was drilled in the MBF A reef, Well #1, which began in September 2016. The well was drilled directionally with a kick-off point at 4305 ft measured depth and was drilled to a total depth of 6085 ft measured depth (5841 ft true vertical depth). A full suite of wireline logs were collected along with sidewall cores and whole cores (Fig. 14). This included triple combo (gamma ray, neutron porosity, density, photoelectric effect, and resistivity), monopole and dipole sonic, deep shear wave, anisotropy, acoustic and resistivity image, elemental spectroscopy, and nuclear magnetic resonance. The collection of whole core from this well was problematic due to the reservoir conditions and directionality of the wellbore. A total of 118.15 ft of whole core was recovered. The cored interval spans a discontinuous section of the A-1 Carbonate and Brown Niagara Formations from 5525 to 5763 ft MD. Drilling occurred between core runs #2 and #3 and between core runs #4 and #5. An additional 69 rotary sidewall cores were collected from multiple formations. The whole core and sidewall core analyses included photographs, descriptions, routine porosity and permeability measurements, and dual-energy CT scans.

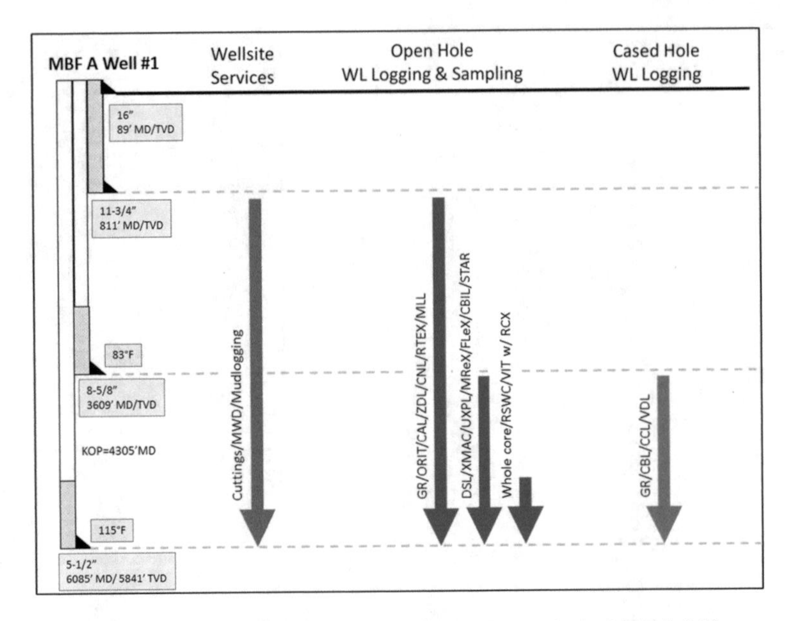

Fig. 14 Data collected from the Well #1 for formation evaluation in the MBF A field

A new well was drilled in the MBF B reef, Well #2, which began in December 2016. The well was drilled directionally from the kickoff point at 4342 ft MD and was drilled to a total depth of 6455 ft MD (6356 ft true vertical depth). A full suite of wireline logs were collected along with sidewall cores and whole core (Fig. 15). This included triple combo, PE, resistivity, monopole and dipole acoustic, deep shear wave, anisotropy, acoustic and resistivity image, and elemental spectroscopy. A total of 210 ft of whole core was recovered over a continuous section in the Brown Niagaran formation from 6,148–6,358 ft MD. Additionally, 30 sidewall cores were collected across multiple formations which were processed for photographs, descriptions, routine porosity and permeability measurements, and dual energy CT scans.

4.3 *Porosity Analysis*

For these wells, the carbonates evaluated were clean with gamma ray values less than 50 API. Following the traditional effective porosity calculation, the original porosity log is equal to the effective porosity because $V_{shale} = 0$. Crossplot porosity was then used to flag specific intervals with relatively higher porosity values, focusing on intervals with porosity greater than 5%. Additionally, water saturation of less than 40% is used to further restrict the porosity zones of interest. In the dolomite case, MBF A, the total porosity was consistently above 5%. There were intervals of higher

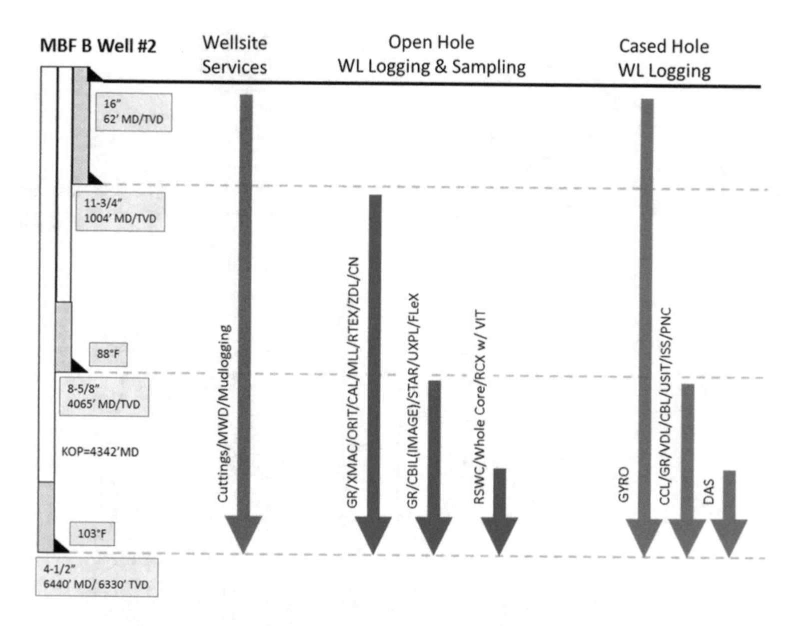

Fig. 15 Data collected from the Well #2 for formation evaluation in the MBF B field

water saturation marking a transitional zone into the waterleg of the reservoir. The limestone case, MBF B, had much lower porosity overall with thin streaks of higher porosity (Fig. 16). The oil water contact (OWC) was more distinct with a strong shift towards 100% water saturation at the base of the reservoir.

4.4 Integration of Log and Core Data

The core collected in the MBF A reef did not correlate with the porosity logs because of challenges in collecting the whole core. Due to this, it was not possible to use for a direct correlation with wireline logs. Instead, the relationship between the core measured porosity and permeability was used to predict continuous permeability for the reservoirs (Fig. 17). The limestone reef MBF B has more success in correlating whole core porosity to wireline log porosity. The resulting correlation was nearly one-one. The relationship between porosity and permeability, however was poor (Fig. 18).

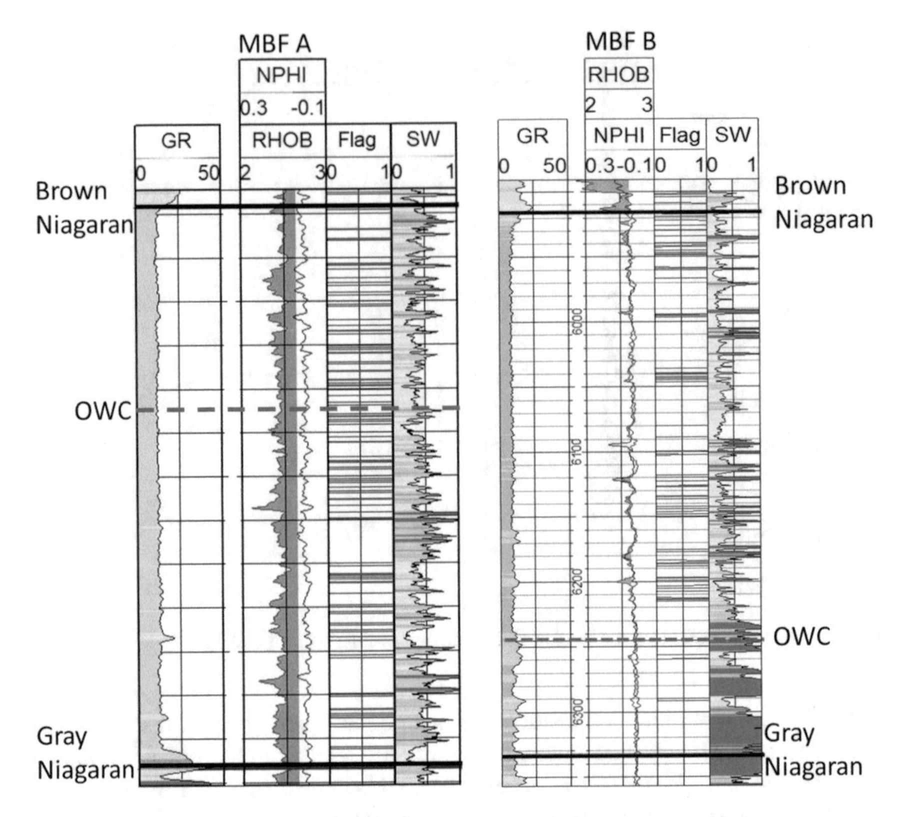

Fig. 16 Example well section of the MBF A well and MBF B well showing reservoir flags (red lines) based on crossplot porosity cut off >5% and water saturation less than 40%

4.5 Facies Analysis

The facies in the two wells were determined by the position of the well within the reef structure, whole core, and wireline log signatures. 3D seismic surveys were available showing that the MBF A well was drilled along the high leeward side of the reef which deviated towards the reef core, or thickest interval of the reef. Wireline logs and whole core confirmed this position by showing low gamma ray and numerous fossils with moldic porosity.

The MBF B well was drilled in the highest portion of the reef, through the main reef core. The wireline logs and whole core confirmed this by showing low gamma ray and evidence of fossils. Figure 19 is an example compilation of whole core photographs, lithology log, and descriptions for the MBF B well.

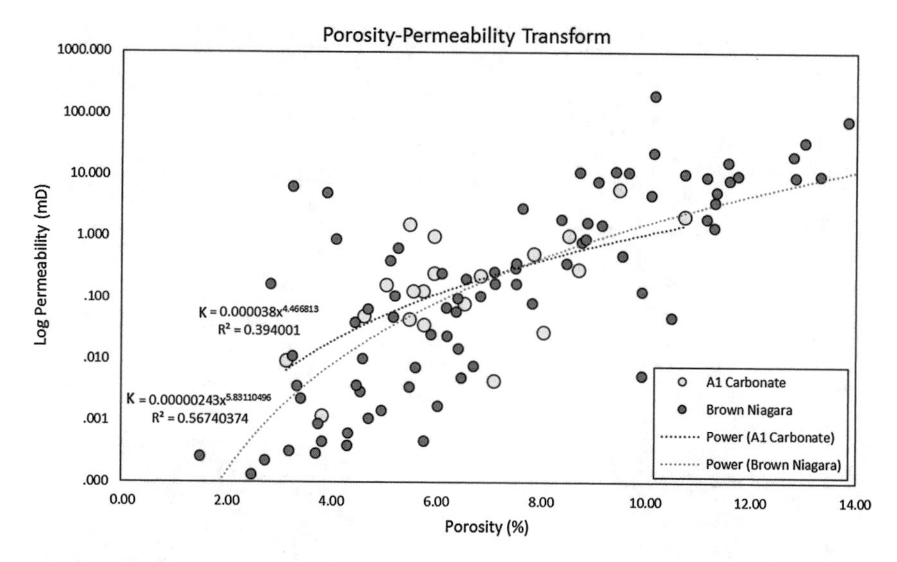

Fig. 17 Core measured porosity and permeability in MBF A for the A-1 Carbonate and Brown Niagaran formations and their resulting transforms

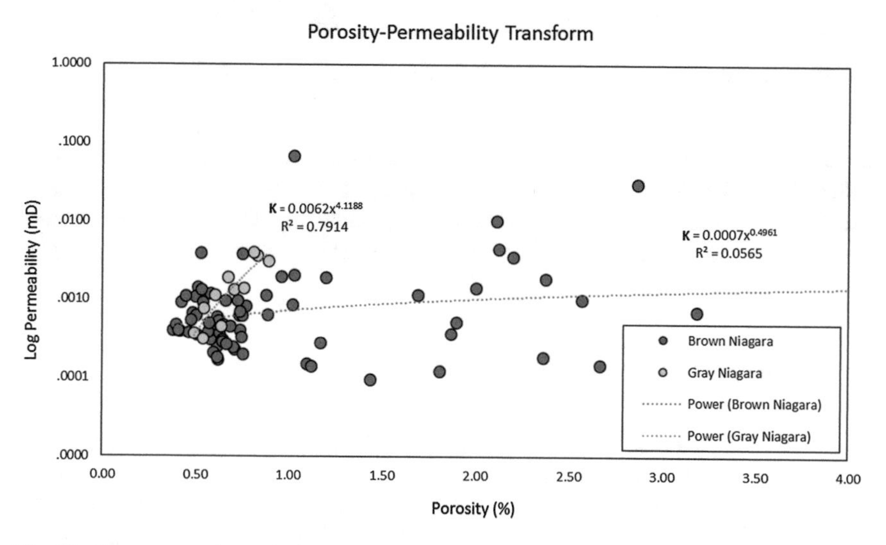

Fig. 18 Core measured porosity and permeability in the MBF B for the Brown Niagaran and Gray Niagaran formations and their resulting transforms

4.6 *Fractures and Vugs*

The whole core collected from the two sites were analyzed using dual energy CT scans to identify and assess the secondary porosity (vugs and fractures) characteristics.

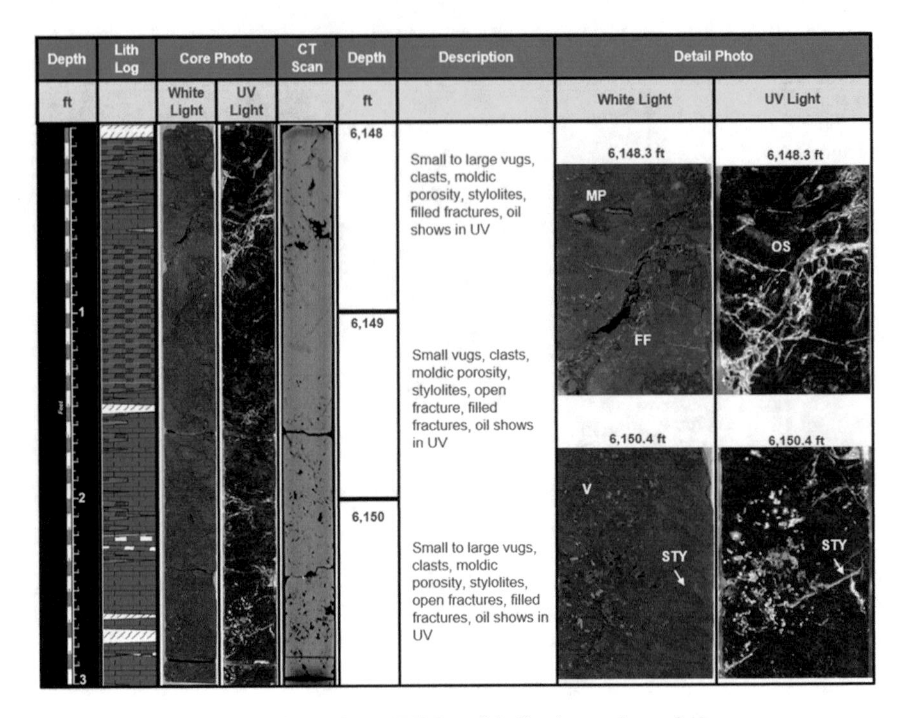

Fig. 19 Whole core descriptions of the MBF B well indicating reef core facies

Thresholds were established to isolate each feature and they were rendered in a 3D space (Fig. 20). The results were later compared to the wireline logs to establish predictive models using machine learning methodologies. Figure 21 demonstrates the results for the MBF A field illustrating identified fractures and salt filled voids.

5 Application of Machine Learning Techniques

There is a longstanding record of the use of statistical techniques for elucidating relationships among geology variables for formation evaluation (e.g., [10]). In recent years, attention has turned to the use of machine learning (ML) techniques for such purposes. ML techniques are algorithms where the functional form of the relationship between independent variables (e.g., log attributes) and dependent variables (e.g., permeability) is not explicitly defined, but is inferred from the data by the algorithm. ML algorithms are also readily adaptable for "big data" problems, where the challenge is (1) acquiring and managing data in large *volumes*, of different *varieties*, and at high *velocities*, and (2) using advanced techniques to "mine" the data and discover hidden patterns of association and relationships in large, complex, multivariate datasets. As noted by Mishra and Datta-Gupta [19], the benefits of such data-driven algorithms can be summarized as: (1) identifying hidden patterns in the

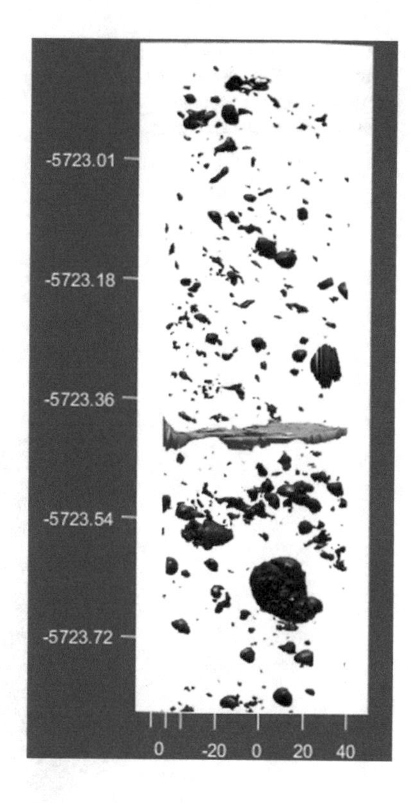

Fig. 20 3D visualization of secondary porosity features isolated in the MBF A whole core. Gray planes represent fractures while red objects represent vugs

data, (2) capturing complex nonlinear relationships, (3) automated learning of the model, and (4) avoiding the need to define full-physics based mathematical relations among input and output variables.

5.1 Machine Learning Basics

The terms "*data mining*", "*machine learning*", "*knowledge discovery*", and "*data analytics*" are generally used interchangeably in this context to denote an exercise where the goal is to extract important patterns and trends, and understand "what the data says", using supervised and/or unsupervised learning [16]. In **supervised learning**, the value of an outcome is predicted based on a number of inputs, with the training data set used to build a predictive model or "learner" via techniques such as conventional or advanced regression analysis methods. On the other hand, **unsupervised learning** involves describing associations/patterns among a set of input measures to understand how the data are organized or clustered, using techniques such as cluster analysis and principal component analysis as well as other methods such as multidimensional scaling and self-organizing maps.

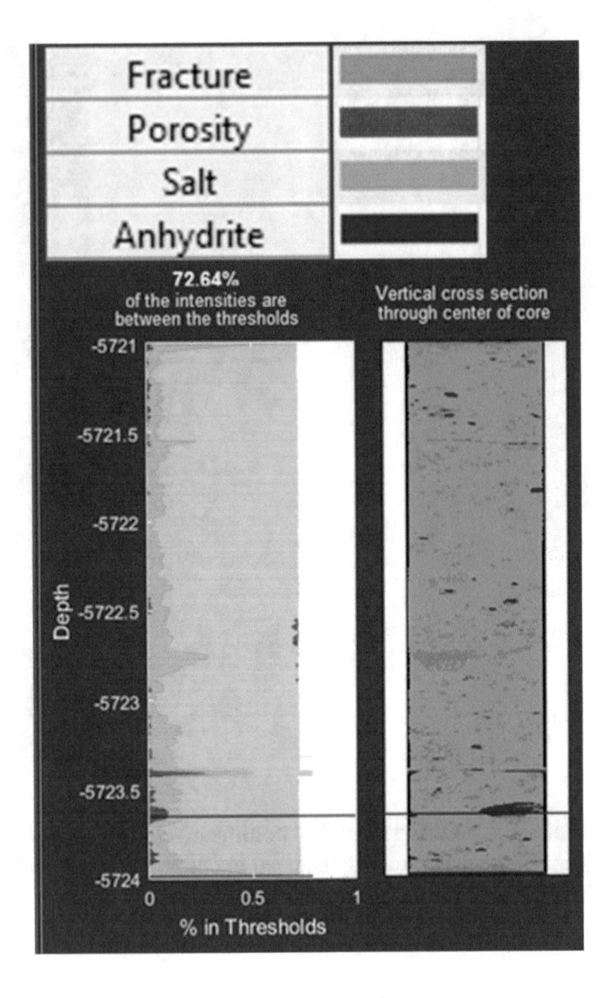

Fig. 21 Example results from the MBF A whole core showing percent secondary porosity, identified fractures, and voids filled with salt

Our focus here is on supervised learning problems, which can be further subdivided into: (a) **regression** problems, where the response variable is continuous, or (b) **classification** problems, where the response variable is categorical. In both cases, the predictor variables can be continuous and/or categorical. For example, building a predictive model for permeability as a function of well-log attributes would be a regression problem [17], whereas determining the factors responsible for identifying electrofacies on the basis of well-log response is a classification problem [21]. Tables 1 and 2 list some of the common ML techniques used for solving regression and classification problems, respectively (see [16] for details).

Representative applications of machine-learning based data-driven modeling for formation evaluation can be found on a number of topics ranging from lithology and facies identification to estimation of rock properties [4, 17, 20–22, 24, 27, 31, 33].

Table 1 Machine learning strategies for regression problems

Random forest regression
• Ensemble of simple regression trees, each of which is trained using random subset of observations and predicators
Gradient boosting machine
• Ensemble of regression trees trained sequentially, with each new tree designed to address shortcomings in predictions made by earlier trees
Support vector regression
• Transforms the data into another space in which a linear regression-style approach can be used to model them
Artificial neural network
• Inputs mapped to outputs via hidden units using a sequence of nonlinear functions of weighted linear combinations of inputs/hidden units
Gaussian process emulation
• Models the response as a trend term with an autocorrelation structure, where neighboring observations have similar responses

Table 2 Machine learning strategies for classification problems

Name	Description
Random forest (RF)	An ensemble of classification trees (see RPART below) that are used to predict the class label of a new observation using a voting scheme
Naïve Bayes (NB)	A classifier based on using Bayes' rule for probability. In this implementation, each class label group is assumed to be Gaussian distributed. A new observation is given the class label of the group for which its sensor logs are most probable based on the distribution
LogitBoost (LB)	An ensemble model created from a collection of "weak" classifiers. In this implementation, the classifiers in the ensemble are decision stumps (single-node decision trees). Classifiers in the ensemble are trained sequentially, where each new classifier is trained to make up for the deficiencies in the previous ones
Gradient boosting machine (GBM)	An ensemble of classification trees. The form of the model is like a random forest model, but the trees are trained sequentially (as in LogitBoost), rather than all at once
Generalized linear model (GLM)	A model where a function of the mean response is modeled as a linear function of the predictors. In this implementation, a logistic function was used
K-nearest neighbor (KNN)	Each new observation is given the class label of the closest training observation, where "closeness" is defined by the Euclidean distance between the vectors of sensor logs

(continued)

Table 2 (continued)

Name	Description
Support vector machine (SVM)	A model that assigns a linear hyperplane in the predictor space to optimally separate two groups. Multi-class SVMs can be constructed by combining several binary classifiers. This implementation uses the "kernel trick" with a Gaussian kernel to generate non-linear boundaries between the groups
Conditional inference tree (CTREE)	A classification tree where, at each node in the tree, a split is made on a predictor. The predictor in this case is selected using the *p*-value of a regression of the response on each individual predictor. The one with the largest association (i.e., the lowest *p*-value) is chosen for the split
Recursive partitioning (RPART)	This methodology, also called "Classification and Regression Trees", or CART, is another way to create decision trees. In this case, the predictor chosen for the split at each node is the one that produces the optimal split according to some criterion. The criterion here was the Gini index, which measures to what degree each group's class labels are diverted down the same branch of the tree

5.2 Example—Vug Characterization

This section describes an application of machine learning techniques for identifying vuggy zones from basic well log attributes using well data from the Appalachian Basin in USA. The presence of vugs in carbonate reservoirs in the study area has often been associated with zones of high production or high injectivity. However, characterization of these zones can be challenging due to the heterogeneity of properties and unpredictable diagenesis. Image logs and whole core can be used to map such zones, but these are not readily available and expensive to collect. Other methods such as using the difference between acoustic porosity and neutron porosity to estimate secondary porosity have not been representative of vugular porosity in the study area due to the neutron porosity logs greatly underestimating the porosity. Machine learning techniques were applied to select wells to develop a vug prediction model from advanced data (whole core, image logs) to readily available wireline logs.

The region of interest has a rich history of oil and gas production and storage making it a key area for deployment of carbon sequestration technologies. The basin is home to numerous point sources of CO_2 emissions, and the region relies heavily on coal-fired power plants for electricity generation. However, many wells focus on shallow plays and there has been little interest in characterizing deep formations for potential storage, such as the carbonates in the Sauk Sequence along the Upper Ohio River valley (Fig. 22).

1. *Geological Background*

This study focused on the sub-Knox, or Sauk sequence, carbonates in the Appalachian Basin region. Sauk carbonate reservoirs generally originated as grain-rich lithofacies,

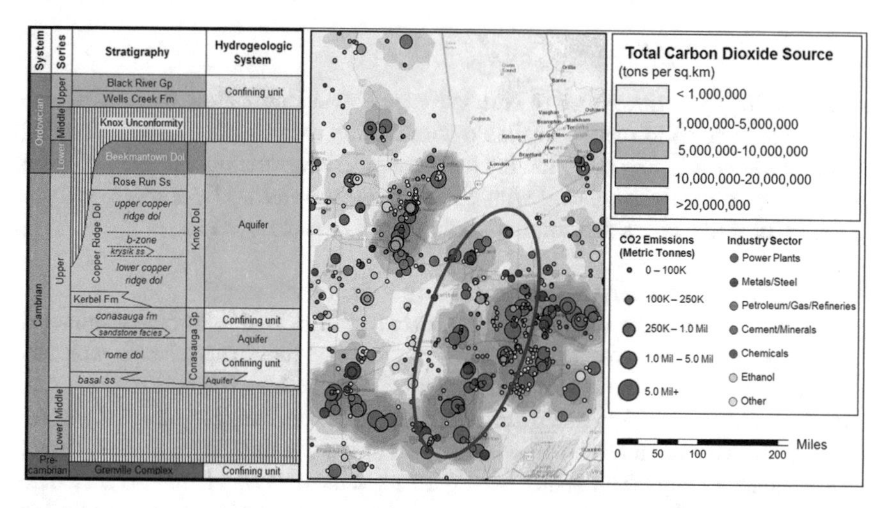

Fig. 22 Stratigraphic column of the Sauk sequence (left) and study area location (right) with CO_2 emissions

deposited in wide facies belts, with primary porosity that was enhanced after burial by dolomitization and other forms of secondary porosity. Oolites and bioclastic skeletal material that collect in shoals and between stromatolites or thrombolites contain original depositional porosity that can subsequently allow burial diagenetic fluids to filter through, often resulting in secondary intercrystalline dolomite porosity or dissolution and vuggy porosity around or in the microbial features, when not occluded by primary micritic or crystalline marine cements [23].

The quality of Sauk carbonate reservoirs within the study area generally depends on the development, preservation, and enhancement of original depositional porosity. Permeability in carbonates is controlled by the degree of pore space connectivity and is related to pore size and shape, abundance of cement and clay, and development of fracture porosity. Within the study area, original porosity and permeability are almost always decreased or occluded after burial, except for the rare development of sucrosic crystalline dolomite from mud-rich carbonates [8]. Thus, reservoir-scale porosity is generally associated with three processes that may involve many stages: (1) primary depositional porosity, enhanced by (2) dolomitization that is then greatly enhanced by (3) subaerial erosion and karst development, post-burial, hydrothermal fluid dissolution, or the development of tectonic fractures.

2. *Methodology and Results*

A methodology was developed to correlate a positive identification of vugs (image logs or whole core) to readily available wireline log data (triple combo). The vug predictor was developing by following two phases: (1) conducting a single well pilot study using a representative well with wireline log data and whole core, (2) using a subset of 10 wells with a full suite of wireline logs as training data.

Phase 1 (selecting best ML algorithm using a single test well)
In the initial phase of the vug analysis, a single well was used to evaluate statistical models that predict the presence of vugs using wireline logs. The selected well had whole core collected along with a full suite of wireline logs (gamma ray, neutron porosity, bulk density, photoelectric effect, resistivity, acoustic, and image logs). An expert geologist examined the whole core and image log and rated, at each depth, the prevalence of vugs. This rating was done on a scale of 0–3, where 0 indicated no vugs present, 1 indicated a light vug density, 2 indicated a medium vug density, and 3 indicated a high vug density. As an example, Fig. 23 illustrates a zone of high vug density and one with no vugs.

A set of models were developed to predict vug prevalence on the basis of wireline logs alone, using the techniques listed earlier in Table 2. The predictors used in the models were a subset of data which included gamma ray (GR), photoelectric (PE), resistivity (RT), density porosity (DPHI), and neutron porosity (NPHI) to be representative of wireline log data typically available in the study area.

The accuracy of each of the models was evaluated using a five-fold cross-validation procedure [19]. Using this approach, the data set was randomly split into five partitions. Each partition is systematically held out, and the model is trained using the other four partitions. The model was then used to make predictions on the held-out observations. After cycling through all of the partitions, this process yielded a single prediction for every observation where the observation was not included in the training set of the model being used. Cross-validated predictions give a more realistic and unbiased evaluation of how well a particular model-building strategy will work when the model is used to make predictions on future independent test data.

The cross-validation procedure was repeated 100 times using different random subsets of the data in order to obtain a more robust measure of the accuracy for the different types of predictive models. The accuracy of the predictions is summarized in Fig. 24. Based on the model comparison, the top performers were the support vector machine (SVM) and gradient boosting machine (GBM) with average correct classification rates of approximately 80%.

Phase 2 (application to a multi-well dataset)
Image logs were used to record a binary vug response: 0 = no vugs present, and 1 = vugs present instead of the four-category 0–3 vug prevalence previously used in phase 1 (so as to simplify the analysis). In order to provide a robust assessment of the classifiers, the logs were down-selected to only those that were present for most

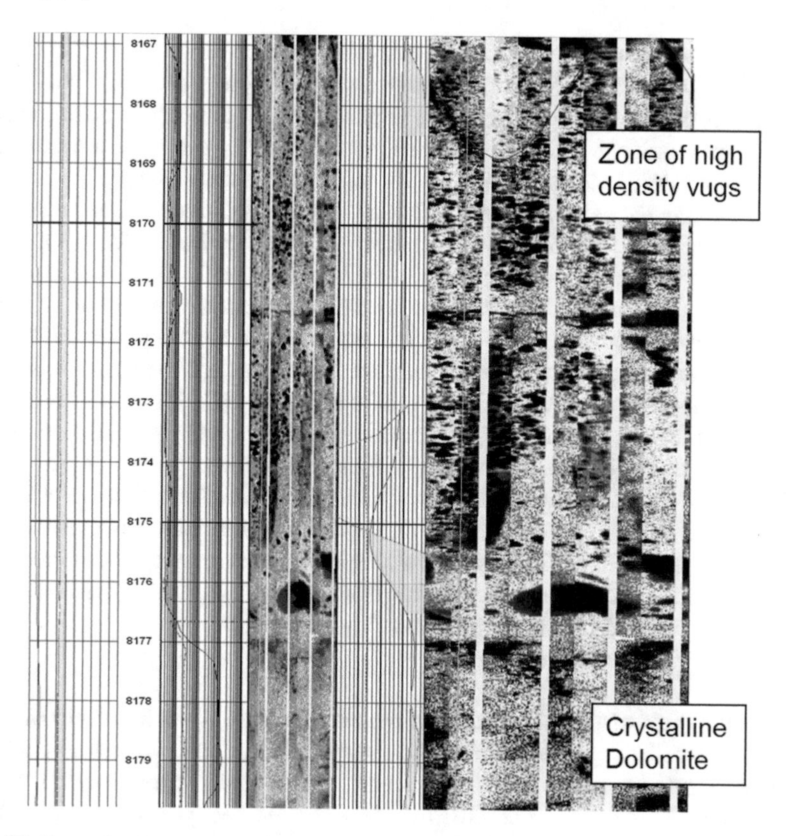

Fig. 23 Example of image log illustrating the upper interval with high-density vugs and the lower interval with no vugs

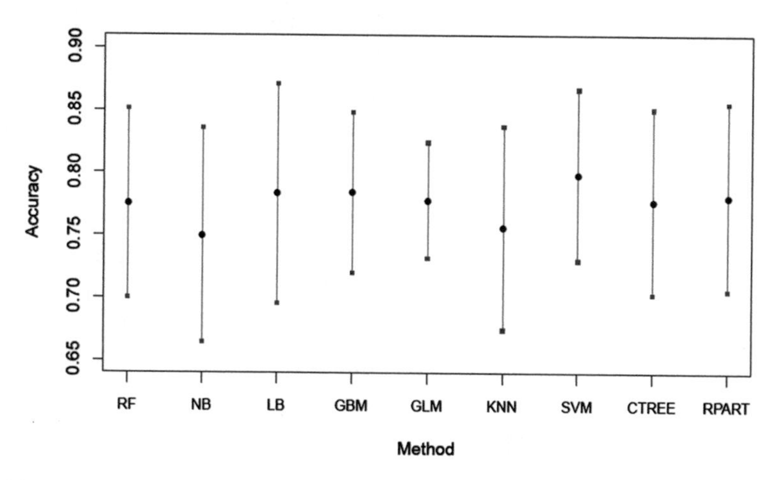

Fig. 24 Prediction accuracy for the statistical models in phase 1. Black dots show the average correct identification rate over the 100 cross-validation runs. The bars extend to ± two standard deviations in the identification rates across the 100 runs

training wells, and only those logs were used to predict vuggy zones. This set of logs is GR (gamma ray), NPHI (neutron porosity), PE (photo-electric), and RHOB (bulk density). Wireline log cutoffs were implemented to eliminate zones of rock not likely to develop vugs by using a gamma ray cutoff of 75 API (shales/mudstone), and a photoelectric index less than 1.81 barnes/electron (sandstones).

Based on the phase I results, the SVM and GBM classifier models were used on this larger set of wells to build classifiers for vug identification. To test the models, they were trained on a subset of wells (Table 3) and the performance was tested on the remaining wells. A well-level cross-validation procedure was used to test each model which allowed every well to play an equal part in the evaluation. Each well was systematically set aside, a vug classifier was trained on the other wells, and then it was used to make a prediction on the held out well. The predictions were compared to the actual binary vug record and the resulting correct identification rate was calculated. Table 3 lists the correct identification rates for each model by well.

The cross-validation results show an overall correct identification rate of about 77% for both modeling approaches, with generally comparable performance on a well-by-well basis. Figure 25 illustrates a comparison between actual identified vugs and the resulting predicted vugs, indicating that zones of high vug incidence centered around 8600 and 8750 ft have been reasonably identified.

3. *Discussion*

The development of the vug prediction model uncovered a wealth of information about how wireline logs can be related to vug prevalence. The pilot phase (phase 1) identified a set of predictive models that can be used to describe this relationship, and showed that the SVM and GBM models had the best predictive ability. The

Table 3 Well-level cross-validation correct identification rates for the SVM and GBM models

Well	Models on subset (XGR < 75, XPE > 1.81)	
	Model 1	Model 2
	SVM	GBM
	XPE, XRHOB, XNPHI, XGR	XPE, XRHOB, XNPHI, XGR
Overall rate	0.7723	0.7704
Well 1	0.7205	0.7077
Well 2	0.6751	0.6866
Well 3	0.7479	0.6482
Well 4	0.8198	0.8097
Well 5	0.7673	0.7973
Well 7	0.8854	0.8839
Well 8	0.7334	0.7366
Well 9	0.6045	0.5718
Well 10	0.8098	0.8023
Well 11	0.8200	0.9012

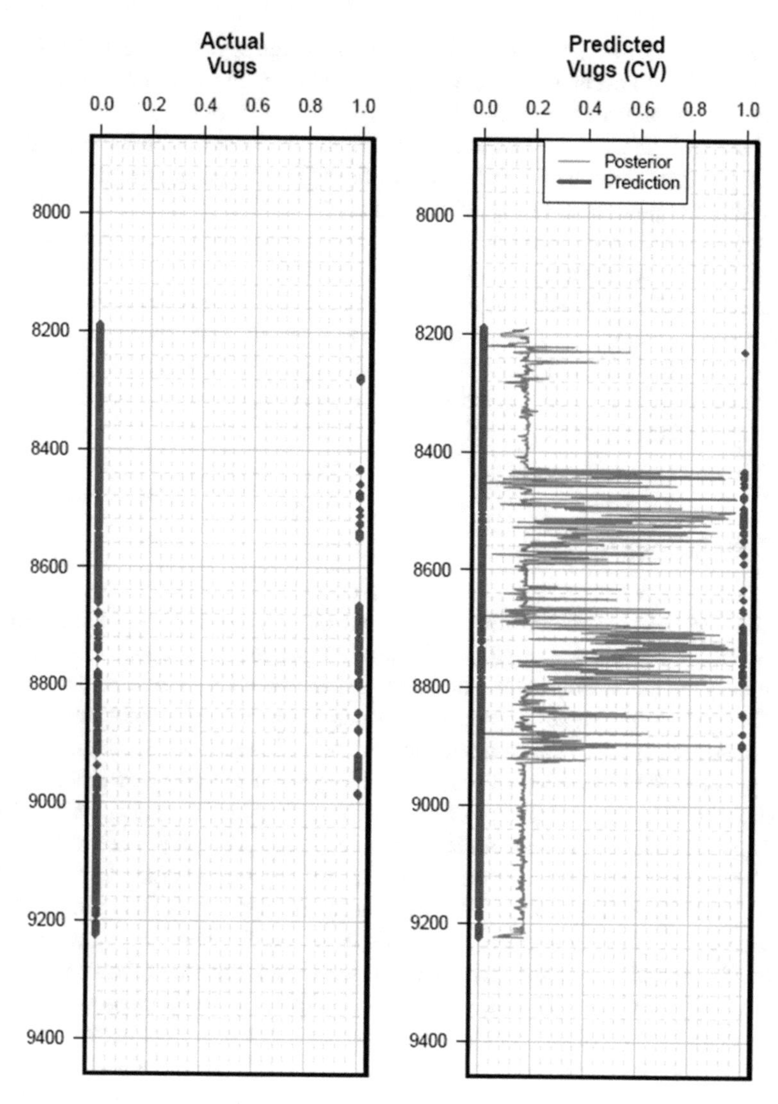

Fig. 25 Actual identified vugs (left) compared to predicted vugs (right) showing high predictability of the vug model

subsequent well comparison study on a multi-well dataset showed on the order of a 60–90% correct identification rate in predicting vugs on wells that had not been used to train the models.

There are several useful outcomes of this study. First, it provides a cross-validation methodology for investigating relationships of this nature that could be applied to other similar problems. Second, it shows that one can, at the very least, identify combinations of wireline log values that are consistent with vug formation. Third, the vug model can be applied to wells without image logs or core to predict the probability of vugs occurring, which was shown to correlate with the injectability of a formation and the concentration of oil in a producing field.

The use of four wireline logs to predict the vugular zones limited the number of wells in the study area which could be ran through the model, specifically the PE log. Additional predictors or log substitutes could be used to expand the dataset to include more wells for a more detailed analysis on a regional scale. Otherwise, the tool was well suited for a local and single well study.

The vug prediction model was very useful in identifying and correlating potential vugular zones, however it cannot differentiate between effective and noneffective vugular porosity. Additional core data is needed to expand the analysis to include a differentiator. Then, the vug prediction model could be used a proxy for permeability, injectivity, and productivity.

6 Concluding Remarks

This chapter seeks to provide a broad overview of well logging in carbonate formations from the perspective of CO_2 sequestration projects. The chapter begins with a summary of the various types of basic and advanced logs that are commonly used (i.e., gamma ray, neutron porosity, bulk density, resistivity, photoelectric, sonic, pulsed neutron capture, nuclear magnetic resonance, image log, and elemental spectroscopy). This is followed by a discussion of the types of analyses performed with the log outputs for characterizing attributes of the formation such as: facies, shale volume, porosity, permeability, fluid saturation, geomechanical properties, and fractures and vugs. Next, the chapter discussed best practices for log interpretation, including data QA/QC and a typical workflow. Based on these concepts, a case study on well-log interpretation is presented, using two fields from the Northern Michigan Pinnacle reef trend—one of which was a limestone reservoir and the other a dolomite reservoir. The case study covers data acquisition, porosity analysis, integration of core and log data, facies analysis, and fractures and vugs. The emerging topic of machine learning is presented next, beginning with a discussion of machine learning basics, which is followed by an example application for vug characterization. Collectively, these topics cover the most common set of conventional and emerging techniques that are relevant in the utilization of well logs for characterization of carbonate formations.

Acknowledgements The research described in this chapter was funded in parts by the United States Department of Energy National Energy Technology Laboratory (NETL) under award #DE-FC26-0NT42589, the Ohio Development Services Agency's Coal Development Office Grant D-15-08.

References

1. Ansari, R., Mekic, N., Chace, D., Rust, M., & Starr, M. (2009). Field applications of a new cased hole gas saturation measurement in tight gas reservoirs. In *Proceedings of the Society of Petrophysicists and Well Log Analysts (SPWLA) 50th Annual Logging Symposium*, Woodland, TX, June 21–24, 2009.
2. Asquith, G., & Krygowski, D. (2004). Basic well log analysis (2nd ed.). In *AAPG Methods in Exploration Series*, No. 16.
3. Bell, K. G. (1963). Uranium in carbonate rocks: A study of ore deposition and occurrence of uranium in carbonate environments, including analyses of 70 samples. In *Shorter Contributions to General Geology, Geological Survey Professional Paper 474-A*. Washington, DC: United States Government Printing Office.
4. Bhattacharya, S., & Mishra, S. (2018). Application of machine learning for facies and fracture prediction using Bayesian Network Theory and Random Forest: Case studies from the Appalachian basin, USA. *Journal of Petroleum Science & Engineering, 170,* 1005–1017.
5. Braunberger, J., Hamling, J., Gorecki, C., Miller, H., Rawson, J., Walsh, F., et al. (2014). Characterization and time-lapse monitoring utilizing pulsed-neutron well logging: Associated CO_2 storage at a commercial CO_2 EOR project. *Energy Procedia, 63,* 3935–3944.
6. Briggs, L. I., Gill, D., Briggs, D. Z., & Elmore, R. D. (1980). Transition from open marine to evaporite deposition in the Silurian Michigan Basin (Chap. 17). In A. Nissenbaum (Ed.), *Hypersaline brines and evaporitic environments: Developments in sedimentology* (Vol. 28, pp. 253–270).
7. Carmichael, R. S. (Ed.). (1982). *Handbook of physical properties of rocks* (Vol. 2, pp. 1–228). Boca Raton, FL: CRC Press Inc.
8. Choquette, P., & Hiatt, E. (2008). Shallow-burial dolomite cement: A major component of many ancient sucrosic dolomites. *Sedimentology, 55,* 423–460.
9. Coates, G., Xiao, L., & Prammer, M. (1999). *NMR logging principles and applications.* Elsevier Science.
10. Davis, J. C. (2002). *Statistics and data analysis in geology.* New York, NY: Wiley.
11. Eyvazzadeh, R., Oscar, K., Hajari, A., Ma, S., & Behair, A. (2004). Modern carbon/oxygen logging methodologies: Comparing hydrocarbon saturation determination techniques. https://doi.org/10.2118/90339-MS
12. Gill, D. (1979). Differential entrapment of oil and gas in Niagara pinnacle-reef belt of northern Michigan. *American Association of Petroleum Geologists, 63*(4), 608–620.
13. Grammer, G. M. (2007). Summary of research through phase II/year 2 of initially approved 3 phase/3 year project—Establishing the relationship between fracture-related dolomite and primary rock fabric on the distribution of reservoirs in the Michigan Basin—Final scientific.
14. Haagsma, A., Rine, M., Sullivan, C., Conner, A., Kelley, M., Modroo, A., et al. (2017). Static earth modeling of diverse Michigan Niagara reefs and implications for CO_2 storage. *Energy Procedia, 114,* 3353–3363.
15. Harrison III, W. B. (2010). EOR in Silurian-Niagaran reef reservoirs, Michigan. In*Evaluation of CO_2-enhanced oil recovery and sequestration opportunities in oil and gas fields in the MRCSP region* (MRCSP Phase II Topical Report), October 2005–October 2010. U.S. DOE Cooperative Agreement No. DE-FC26-05NT42589, OCDO Grant Agreement No. DC-05-13 (pp. 48–54).
16. Hastie, T., Tibshirani, R., & Friedman, J. H. (2008). *The elements of statistical learning: Data mining, inference, and prediction.* New York, NY: Springer.
17. Lee, S., Kharghoria, A., & Datta-Gupta, A. (2002). Electrofacies characterization and permeability predictions in complex reservoirs. *Society of Petroleum Engineers, Reservoir Evaluation and Engineering,* 237–248.
18. Li, K., & Horne, R. (2005). Inferring relative permeability from resistivity well logging. In *Thirtieth Workshop on Geothermal Reservoir Engineering*, Stanford University, Stanford, CA, January 31–February 2, 2005.

19. Mishra, S., & Datta-Gupta, A. (2017). *Applied statistical modeling and data analytics*. Elsevier.
20. Mishra, S., Li, H., & He, J. (2019). *Machine learning in subsurface characterization*. Houston, TX: Gulf Professional Publishing.
21. Perez, H., Gupta, A. D., & Mishra, S. (2005). Hydraulic flow units in permeability prediction from well logs: A comparative analysis using classification trees. *Society of Petroleum Engineers, Reservoir Evaluation and Engineering,* 143–155.
22. Qi, L. S., & Carr, T. R. (2006). Neural network prediction of carbonate lithofacies from well logs, Big Bow and Sand Arroyo Creek fields, Southwest Kansas. *Computers & Geosciences, 32,* 947–964.
23. Ryder, R. T. (1994). The Knox unconformity and adjoining strata, western Morrow County, Ohio. In W. E. Shafer (Ed.), *The Morrow County, Ohio "oil boom" 1961–1967 and the Cambro-Ordovician reservoir of central Ohio* (pp. 249–271). Columbus: Ohio Geological Society.
24. Salehi, S. M., & Bizhan, H. (2014). Automatic identification of formation lithology from well log data: A machine learning approach. *Journal of Petroleum Science Research, 3*(2), 73–82.
25. Sarg, J. F. (2001). The sequence stratigraphy, sedimentology, and economic importance of evaporite-carbonate transitions: A review: *Sedimentary Geology, 140*, 9–42.
26. Sears, S., & Lucia, F. (1979). Reef-growth model for Silurian pinnacle reefs, northern Michigan reef trend. *Geology, 7*, 299–302.
27. Sebtosheikh, M. A., & Salehi, A. (2015). Lithology prediction by support vector classifiers using inverted seismic attributes data and petrophysical logs as a new approach and investigation of training data set size effect on its performance in a heterogeneous carbonate reservoir. *Journal of Petroleum Science and Engineering, 134,* 143–149.
28. Shalaby, M. R., & Islam, M. A. (2017). Fracture detections using conventional well logging in carbonate Matulla Formation, Geisum oil field, southern Gulf of Suez, Egypt. *Journal of Petroleum Exploration Production Technology, 7,* 977–989.
29. Shouldice, J. R. (1955). Silurian reefs of southwestern Ontario. *Canadian Mineral and Metallurgy Bulletin, 48*(520).
30. Slatt, R. M. (2006). Stratigraphic reservoir characterization for petroleum geologists, geophysicists and engineers. In *Handbook of petroleum exploration and production* (Vol. 6).
31. Wang, G., Carr, T. R., Ju, Y., & Li, C. (2014). Identifying organic-rich Marcellus Shale lithofacies by support vector machine classifier in the Appalachian basin. *Computers & Geosciences, 64,* 52–60.
32. Westphal, H., Surholt, I., Kiesl, C., Thern, H., & Kruspe, T. (2005). NMR measurements in carbonate rocks: Problems and an approach to a solution. *Pure and Applied Geophysics, 162,* 549–570. https://doi.org/10.1007/s00024-004-2621-3
33. Wu, W., Grana, D., Campbell-Stone, E., & McLaughlin, F. (2015). Bayesian facies classification in a CO_2 sequestration study using statistical rock physics modeling of elastic and electrical properties. In *Presented at the SEG Annual Meeting,* New Orleans.
34. Xiao, L., & Li, K. (2011). Characteristics of the nuclear magnetic resonance logging response in fracture oil and gas reservoirs. *New Journal of Physics.* https://doi.org/10.1088/1367-2630/13/4/045003

Laboratory Scale Works

Alberto Ramos, Carlos Martínez, and J. Carlos de Dios

Abstract Site characterization for CO_2 geological storage needs works at field scale and laboratory to prove the seal-reservoir pair capability to trapping the captured gas in a safely and efficient manner. Laboratory works are used to design field tests and for checking their results in a more controllable monitored environment. Petrophysical characterization through laboratory works intends to quantify operating parameters such as the reservoir injectivity and also the short and long-term trapping ability. For this, both usual routine tests of oil and gas industry, as well as, specific and innovative tests for CO_2 geological storage are carried out. This chapter address laboratory tests and results achieved during the characterization of Hontomín Technology Development Plant (TDP). The main singularity is the poor primary permeability of the naturally fractured carbonates that compose the reservoir, which impacts on gas migration that is dominated by fractures. This challenging fact conditions the petrophysical characterization, since hydrodynamic, geomechanical and chemical effects take place induced by the injection and they must be analyzed by innovative works. Innovative tests and results are described, analyzed and discussed, paying special attention to those carried out by the ATAP equipment, a patent of Technical University of Madrid (UPM).

Keywords Petrophysical characterization · CO_2 geological storage · Injectivity and trapping ability · Hontomín TDP · Routine tests · ATAP test

---

A. Ramos (✉) · C. Martínez
School of Mines and Energy, Technical University of Madrid, Calle de Rios Rosas 21, 28003 Madrid, Spain
e-mail: alberto.ramos@upm.es

J. C. de Dios
Foundation Ciudad de la Energía-CIUDEN F.S.P., Avenida del Presidente Rodríguez Zapatero, 24492 Cubillos del Sil, Spain

J. C. de Dios et al. (eds.), *CO_2 Injection in the Network of Carbonate Fractures*, Petroleum Engineering, https://doi.org/10.1007/978-3-030-62986-1_3

1 Introduction

Site characterization for CO_2 geological storage requires laboratory scale works as other activities such as oil and gas [1] or mining exploration [2], in order to determine main rock properties that support the viability of the project. Positive results from lab tests do not assure a successfully venture, but negative results may jeopardize the project.

During the exploration phase, petrophysical lab results are correlated with inputs from well logging [3] gained during drilling, hydraulic characterization works [4] (see Chapter "On-Site Hydraulic Characterization Tests") and provide data for geological modeling [5]. Lab results are used to perform the injection of well fluids to verify the reservoir behavior and design efficient injection strategies that preserve the integrity of seal formation.

Operating parameters such as injectivity and trapping mechanisms may be determined by lab tests. For this, it is necessary to develop both routine and specific tests. Routine tests are the usual ones that are carried out in the exploration works of oil and gas industry to determine the petrological and petrophysical properties of the cap-rock and reservoir. Their results define the necessary conditions of competency for the formations intend to be the seal and reservoir respectively. Specific tests injecting artificial brine and CO_2 under reservoir condition are necessary to prove the reservoir ability and seal integrity for CO_2 trapping in safely and efficient manner.

Impurities in the captured CO_2 stream [6] upset the resident brine geochemistry of reservoir, impacting the reactivity transport in the carbonate fracture [7]. These effects may vary the conditions of injection and trapping, what means a dynamic behavior of seal-reservoir pair due to the change of their hydraulic properties along the project life. This relevant issue must be tackled by specific lab works.

This chapter address the petrophysical lab works conducted during the hydraulic characterization and early injection at Hontomín Technology Development Plant, describing, analyzing and discussing the results gained. Petrological and petrophysical properties of Marly Liassic as main seal and Sopeña Formation, a naturally fractured carbonate with poor primary permeability as reservoir, are shown, as well as, the results from the co-injection of brine and CO_2 under reservoir condition using ATAP equipment. These results were used as inputs to design field injection tests (see Chapter "On-Site Hydraulic Characterization Tests"). Short-term effects of impurities in the carbonated formations and the preliminary trapping capacity determined by specific lab works are also addressed in the chapter.

2 Petrophysical Characterization Goals

Main goal of petrophysical lab works is to provide realistic information on geological formations planned to be the seal-reservoir pair of the site for CO_2 geological storage.

Therefore, the quantity and representativeness of samples play a key role on this matter, as will be analyzed in the next section.

Thus, lab tests are intended to achieve the following goals:

1. Determine injectivity and storage ability
2. Assess the degree of CO_2 trapping in short-long term.

Injectivity and storage ability are operating parameters that condition the project feasibility. On the other hand, assure the long-term CO_2 trapping is the necessary condition for a positive decision making about the viability of the project.

As far as possible, results from lab works must be contrasted by field scale tests to prove their representativeness about the site properties. Therefore, these works should be developed along the project life, i.e. during the phases of characterization, operation, monitoring previously closure and abandonment.

The study of the facies which compound the seal-reservoir pair needs petrological tests as well, to determine the texture, chemical and mineralogical composition. Moreover, petrophysical properties (e.g. density, porosity, porous distribution and permeability) must be determined. Particularly, geomechanics (e.g. uniaxial strength, tensile strength, Young modulus, Poisson coefficient, cohesion and internal friction angle) provide relevant information about the stress-strain behavior of geological complex under injection condition.

Results from facies study are used for geological modeling, including input from well logging, particularly, from acoustic televiewer [8] to analyze the spatial fracture distribution. Figure 1 shows the facies, televiewer and effective porosity logs conducted below 1000 m MD at the injection (HI) and observation (HA) wells of Hontomín pilot used to build up the static model [5].

Through lab injection tests performed under reservoir condition, i.e. high pressure and high temperature, it is intended to analyze the hydrodynamic, mechanical and chemical effects of biphasic flow CO_2 + brine on rock matrix and fractures. Their results are used to design field characterization tests (see Chapter "On-Site Hydraulic Characterization Tests") and safe and efficient injection strategies (see Chapter "Safe and Efficient CO_2 Injection").

Finally, impurities in the CO_2 stream impact on operation efficiency and storage safety. Regarding the efficiency impacts, some examples may be found at the work developed by de Dios et al. [7], where the effects of N_2 and O_2 in CO_2 stream induce the increase of injection pressure and the decrease of stored CO_2 density in the reservoir. These effects were checked by field scale tests. Storage safety impacts are related to the chemical reactivity between the impurities, CO_2 and resident brine in the reservoir. These effects are analyzed at lab scale in a more controllable monitored environment. Some results are also shown at de Dios et al. [7].

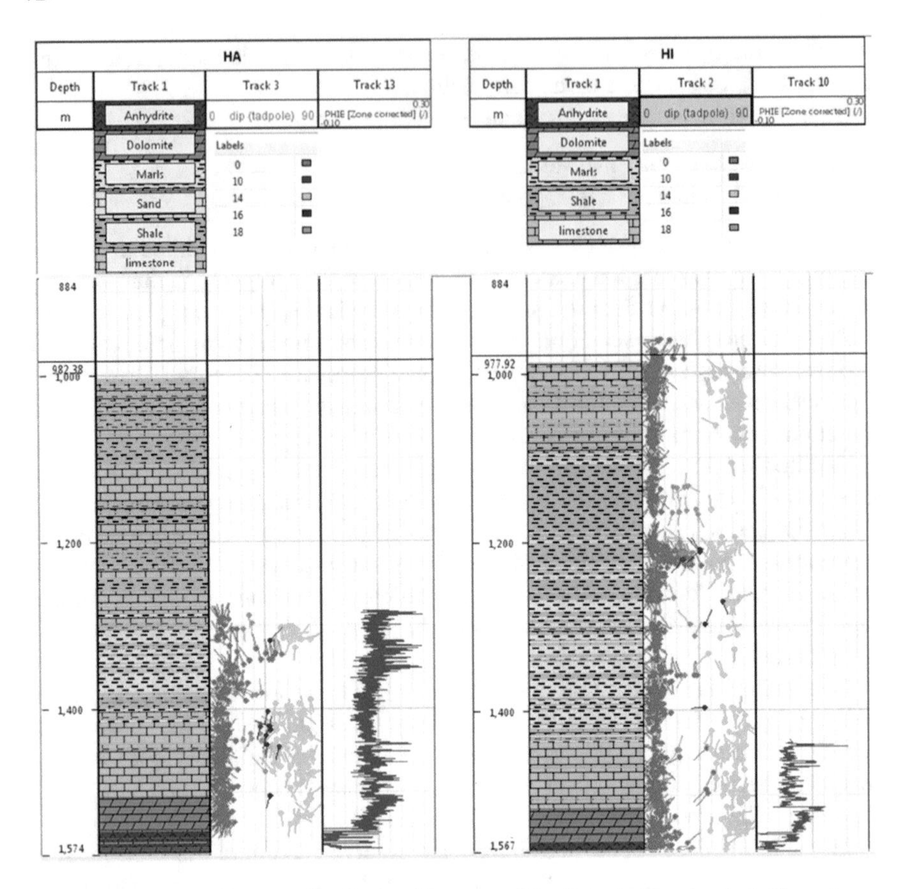

Fig. 1 Facies, spatial fracture distribution (Courtesy of Instituto de Ciencias de la Tierra Jaume Almera) and porosity logs of Hontomín seal-reservoir pair

3 Sampling Methodology

As mentioned above, the representativeness of samples plays a key role to achieve realistic results which may help for a successfully decision making on project feasibility and provide valuable data to design field scale tests and safe injection strategies.

Outcrops [9] may be used to gain initial results in the first stage of site characterization phase, which must be contrasted with those achieved using core samples to reduce the uncertainty in the seal-reservoir exploration by providing data representative in situ conditions. Test typology and results analyzed hereinafter are focused on lab works carried out using core samples extracted during well drilling.

As described in Sect. 2.5 of Chapter "Light Drilling, Well Completion and Deep Monitoring", 10 core samples were extracted during the observation well (HA) drilling and 3 core samples from the injection well (HI). Figure 2 of Chapter "Light

Drilling, Well Completion and Deep Monitoring" shows the well schemes and the geological formations crossed during drilling. Table 1 indicates sampling location in both wells.

7 core samples correspond to the cap-rock (Marly Lias Formation) and 6 to reservoir (4 from Limestone and 2 from Dolomitic Sopeña Formations). Figures 2 and 3 show HA sample number 10 extraction, which corresponds to Dolomitic Sopeña Formation.

Every core sample was scanned using 16-slice Helical CT Scanner, achieving a 2D longitudinal reconstructed tomogram image. The images were done each 1 mm with a total of 1000 images per meter of core sample. Figure 4 shows the Computed Tomography of HA sample number 10.

Subsequently, 49 plugs (1.5″ø, 40 mm L) were extracted parallel to drilling axis (27 units) for geomechanics and perpendicular (22 units) for permeability tests. Plugs were washed firstly with toluene and subsequently with methanol, and finally dried on the stove (50–60 °C). Figure 5 shows plug HP5V used for permeability tests.

Table 1 Location of core samples in HA and HI wells

HA well core sample	Recovery		Drilling interval (m)
	Length drilled (m)	%	
1	7	100	1307–1314
2	3.5	100	1320–1323.5
3	6	85.7	1343–1349
4	4	100	1401–1405
5	5.10	72.8	1405–1410
6	6.77	97	1442–1449
7	1.38	100	1449–1450.38
8	5.87	0	1457–1462.87
9	0.12	60	1464–1462.1
10	6.91	98.7	1515–1522
HI well core sample	Recovery		Drilling interval (m)
	Length drilled (m)	%	
1	7	100	1355–1362
2	0.96	19	1467.74–1468.7
3	6.96	99.12	1531–1538

Fig. 2 Core sample extraction (Courtesy of CIUDEN)

Fig. 3 HA sample 10 (drilling interval: 1515–1522 m) (Courtesy of CIUDEN)

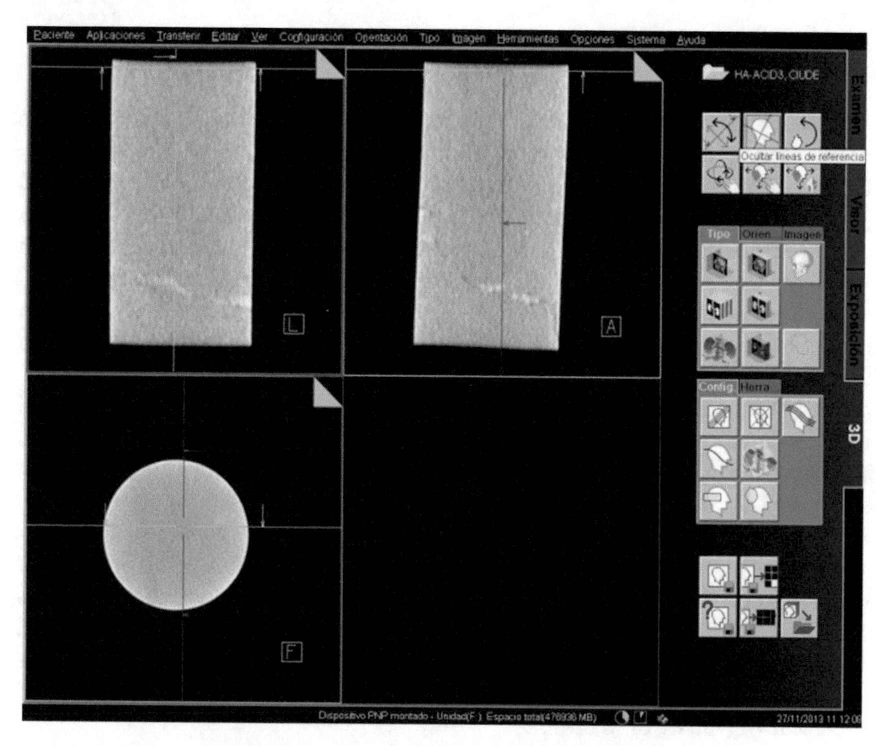

Fig. 4 Computed tomography of HA 10 core sample

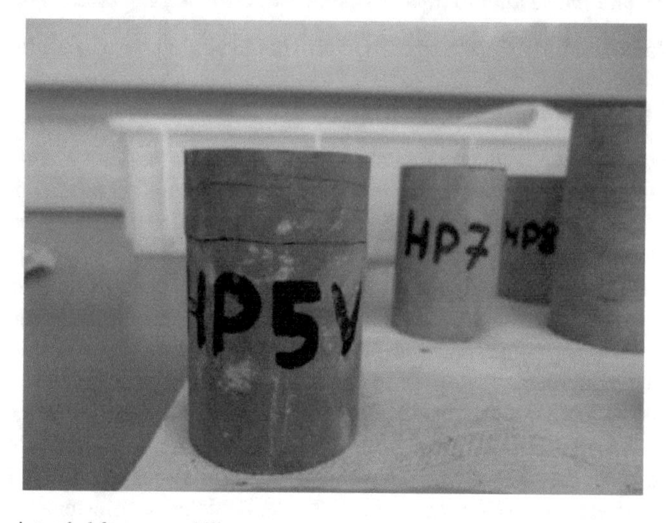

Fig. 5 Plug intended for permeability test

4 Tests to Determine Injectivity and Trapping Mechanisms

As mentioned above, main goal of lab works is to determine CO_2 injectivity [10] in reservoir samples and trapping mechanisms [11] of the geological complex. Injectivity is an operating parameter defined as the rate between the flow injected and pressure needed. Operation effectiveness is conditioned by injectivity, as a high parameter value means higher CO_2 migration than to lower value. Moreover, high injectivity also means safer injection since lower pressure is required.

Both routine and specific tests are necessary to determine injectivity. Routine tests are the usual carried out during geological exploration works (e.g. Oil and Gas, mining and hydrogeology, inter alia activities) to determine petrological and petrophysical properties of seal-reservoir pair. Specific lab tests are performed injecting fluids in reservoir samples under reservoir condition (pressure/temperature) to analyze the hydrodynamic, mechanical and chemical impacts of CO_2 injection in carbonate fractures [4]. Results from lab tests are used as modeling inputs to estimate the storage capacity [11].

In order to assess the dynamic behavior of seal-reservoir pair, the equipment required to carry out specific tests must be able to perform and monitor injections under hydraulic and thermodynamic conditions existing in the geological complex. The equipment used to carry out this type of tests on samples from Hontomín geological formations is ATAP "Equipment for petrophysical tests" P201231913 patent of Technical University of Madrid.

ATAP equipment [12] consists of two ISCO pumps, one of them to keep CO_2 in dense phase and the second to inject brine in the following injection conditions:

- Up to 75 bar and 31 °C to keep carbon dioxide in dense phase in contact with the rock matrix
- Confinement pressure up to 500 bar.

Co-injection of CO_2 and brine can be made on rock samples maintaining the reservoir condition of pressure and temperature. ATAP operating modes are dynamic and static. The first one is used to analyze the effects related to the fluid injection in the rock sample and with the second the effects associated with the trapping evolution. ATAP results are used to study the relative permeability of co-injections and analyze the hydrodynamic, mechanical and chemical effects produced in the sample, such as porosity changes, mineralization processes and trapping mechanisms. Figure 6 shows ATAP equipment.

On the other hand, long-term trapping assessment needs from lab works to determine the evolution of petrophysical properties and perform the resident brine analysis along the project life. These works have not been carried out yet at Hontomín pilot because of the early injection stage of the site. More detailed information about this matter may be found at de Dios et al. [13].

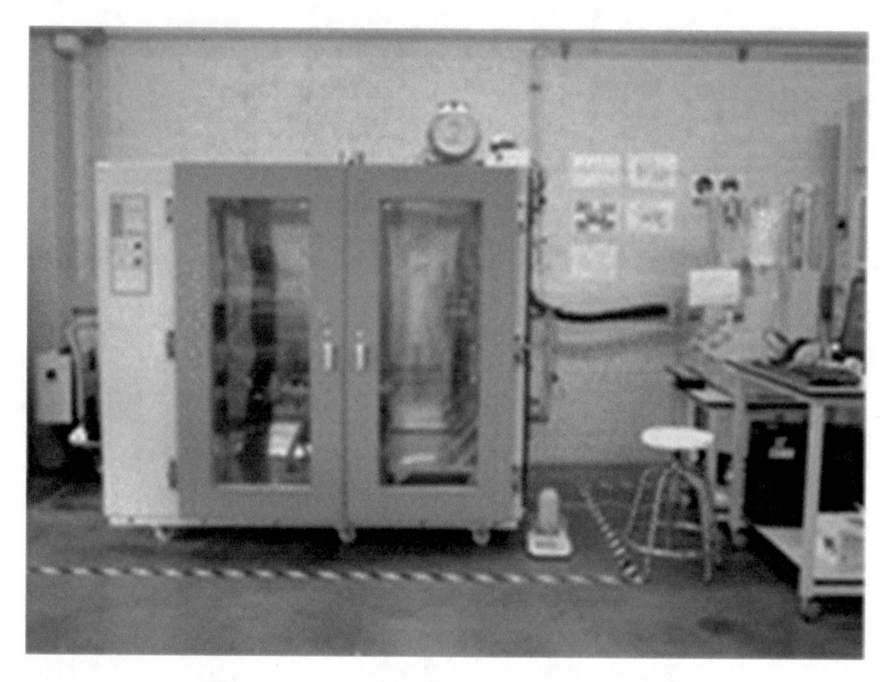

Fig. 6 ATAP equipment (Courtesy of Universidad Politécnica de Madrid)

4.1 Routine Tests

Routine or conventional core analysis usually involves fluid saturation and petrophysical measurements carried out on samples at laboratory conditions. The results are used to characterize the properties of seal-reservoir pair [14].

Lab core logging is typically included as part of routine tests, whose data added to those from the rest of tests and well logging are used for log-core integration [15]. This is the conventional way to characterize the geological formations during the exploration phase of oil and gas and mining projects, inter alia.

Lab core logging carried out on Hontomín samples was:

- Gamma ray core logging
- Sonic core logging
- Magnetic susceptibility core logging
- Electric resistivity core logging.

Figure 7 shows gamma ray core logging performed on Hontomín HI core sample 3 (reservoir).

Main routine tests carried out on core samples were the following:

- Petrology (Texture, chemical and mineralogical composition)
- Petrophysics (Density, porosity, porous distribution, absolute/relative permeability)

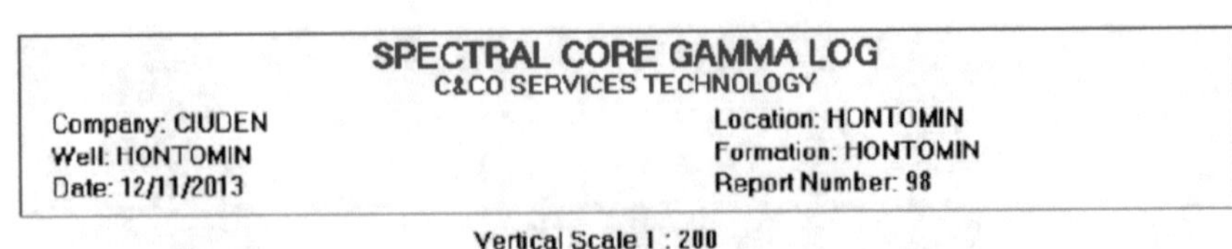

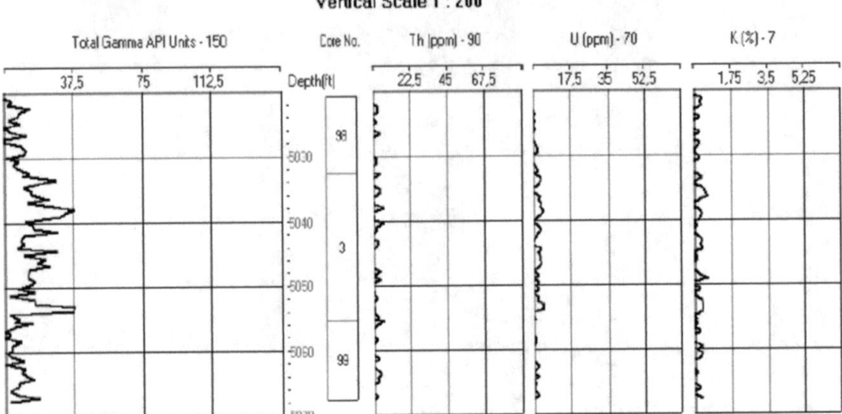

Fig. 7 Hontomín reservoir gamma ray core logging

- Geomechanics (Uniaxial and tensile strength, Young modulus, Poisson coefficient, cohesion and internal friction angle).

4.1.1 Petrological Properties

Petrological studies were focused on formation processes of the rocks that compound the seal-reservoir pair of Hontomín pilot, and particularly, the physic, chemical, mineralogical and spatial properties of geological formations of which they are part. Thus, rock texture, chemical and mineralogical composition are the main petrological properties to define.

Scanning Electron Microscopy (SEM) was used to achieve high resolution images of plug surface, and in this way, to find out the pore size and shape in the rock matrix, the alternation of minerals in the sample, the existing heterogeneities and the fracture characteristics, particularly the nature of the fillings. SEM and CT scanner are techniques used prior and subsequently of injection tests under reservoir condition, in order to analyze texture changes induced in the samples, as will be described in Sect. 4.2. Figure 8 shows SEM image of a reservoir sample, with a filled fracture.

To analyze the microtexture, mineral content and structure, thin and polished sheets were used. Figure 9 shows microscopic image of Dolomitic Sopeña sample using these techniques [16].

X-Ray Fluorescence (XRF) was used on pulverized and compacted core rocks, with a precision greater than the standard deviation of the actual composition of the sample. Rock chemical composition plays a key role to analyze the interaction

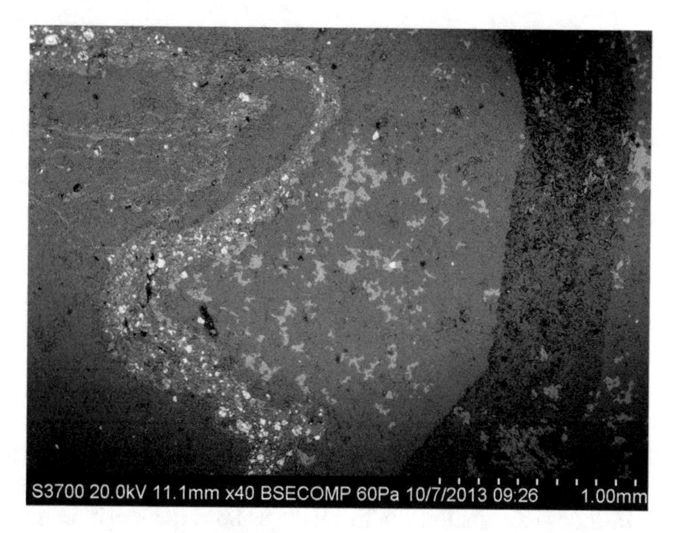

Fig. 8 SEM image of Dolomitic Sopeña plug

Fig. 9 Microscopic image of Dolomitic Sopeña sample

between CO_2, resident brine and the matrix and fracture fillings. Therefore, XRF technique may be used prior and subsequently of CO_2 injections into the plugs, supported by data from chromatographic analysis of effluent, in order to quantify the geochemical reactivity degree. Mineralogical composition of Hontomín seal-reservoir pair was determined using X-Ray Diffraction (XRD), being analyzed both qualitatively and quantitatively. These data were correlated with those coming from the analysis of cuttings during well drilling, and used to define the facies of the site. Table 2 shows XRD data from Hontomín reservoir core samples [16].

Table 2 XRD results from Hontomín reservoir samples

Depth (m)	Mineralogical composition
1443.1	Dolomite 99%, quartz 1%
1444.95	Calcite 49%, dolomite 50%, quartz 1%
1447.55	Calcite 90%, dolomite 9%, quartz 1%
1447.62	Calcite 45%, dolomite 30%, mica 9%, quartz 8%, celestine 8%
1449	Calcite 90%, dolomite 8%, quartz 2%
1450.34	Dolomite 50%, calcite 40%, quartz 8%, mica 2%

Marly Liassic, as Hontomín main seal, is composed by alternating sequence of marls, marly limestones, calcareous mudstones and shales, with the following mineralogical composition: calcite (50–80%), mica (biotite, muscovite), chlorite, quartz, pyrite. Limestone Sopeña, as upper reservoir, is composed by mudstone/wackestone, dolomite grainstone intercalations, with heterogeneous composition depending on the layer analyzed (90% calcite—99% dolomite). And finally, Dolomitic Sopeña composed by pure dolomitic grainstone and increasing anhydrite content because of the proximity of Keuper Formation (Carniolas) (see Fig. 1 of Chapter "Light Drilling, Well Completion and Deep Monitoring").

4.1.2 Petrophysical Properties

Petrophysical characteristics of geological formations condition their ability to become seal-reservoir complex for safely trapping of injected CO_2. Moreover, CO_2 solubility in the resident brine produces chemical reactivity in carbonates that likely induces changes of petrophysical properties [17], such as the primary porosity in the rock matrix and the secondary in fractures.

Main petrophysical properties studied on Hontomín seal-reservoir samples were the following:

- Density
- Porosity
- Pore size distribution
- Absolute/relative permeability.

Density of samples from the seal and reservoir formations was determined using lab scales to measure mass and helium pycnometer to measure volume. It was also determined the moisture of each sample. Density data were correlated with lab gamma ray core logging and also with well logging results. Table 3 shows average density values of Hontomín complex formations.

Open porosity was determined using the helium porosimeter [18] and mercury pycnometer [19]. Through the first test, a specific volume of helium contained in a reference cell is slowly pressurized and then isothermally expanded into the empty volume in the pore matrix. After expansion, the resulting equilibrium pressure will

Table 3 Density of Hontomín complex formations

Geological formation	Average density (g/cm^3)
Marly Liassic (cap-rock)	2.6
Limestone Sopeña (reservoir)	2.7
Dolomitic Sopeña (reservoir)	2.8

be determined by the magnitude of the volume. Using this value and Boyle's Law, the pore volume of the sample is established. Table 4 shows average porosity values of Hontomín seal-reservoir pair.

Marly Liassic data are accordingly to those corresponding to cap-rock formations, but those corresponding to Sopeña show poor values for a reservoir. This is the first clue that lead us to think the CO_2 migration is dominated by fractures in this dual medium.

Mercury porosimetry test [19] is intended to determine pore throat radius and pore throat size distribution. Thus, the test provides data to set the percentage of macro, meso and microporosity in the samples. Test is conducted in two steps, first one in low pressure (0–30 psi) and second in high pressure (30–60,000 psi) injecting mercury into the sample. The equipment can measure the volume entering the sample for each pressure step applied and the volume of mercury remaining. In this way, the volume of mercury trapped in the pore network of sample can be determined at the end of the test. Table 5 shows the pore throat size distribution from samples of Hontomín seal-reservoir pair.

Samples from Marly Liassic show pore throat radius smaller than 0.5 μm in 83–97% of cases, with macroporosity equal or smaller than 9%, being mostly microporosity. Limestone Sopeña samples show greater pore throat radius than 0.5 μm and higher macroporosity (23%) in some cases, than those corresponding to Marly Liassic. Finally, Dolomitic Sopeña samples show pore throat radius and porosity values closer to Marly Liassic data.

Figure 10 shows the Pore Throat Radius Distribution (PTRD) versus pore radius using mercury porosimetry data from Limestone Sopeña samples (HP10, HP12 and HP13). Micro, meso and macroporosity ranges are also plot in the graph, being (0.001–0.5 μm), (0.5–1.5 μm) and ($\geq$0.5 μm) respectively.

Graph shows the porosity of the analyzed rock samples is low, with pore diameters between 1 and 10 μm, typical of reservoirs in which there is dual porosity with low value in the matrix.

The nitrogen adsorption method [20] is a non-destructive technique that allows to characterize the effective micro and mesoporosity of the rock sample. This test

Table 4 Open porosity of Hontomín complex formations

Geological formation	$\o_{open}$ (%)
Marly Liassic (cap-rock)	From 2 to 4
Limestone Sopeña (reservoir)	From 0.2 to 16
Dolomitic Sopeña (reservoir)	From 0.5 to 12

Table 5 Macro, meso and microporosity of Hontomín seal-reservoir pair

Pore throat size distribution			
Geological formation	Macroporosity (%)	Mesoporosity (%)	Microporosity (%)
Marly Liassic	8.09	0	91.91
	1.83	0	98.17
	8.96	1.55	89.49
	7.55	1.26	91.18
	4.15	0	95.85
	2.58	0.14	97.28
Limestone Sopeña	2.47	60	37.53
	23.08	3.20	73.72
	20.73	1.49	77.78
Dolomitic Sopeña	9.74	3.45	86.81
	5.87	1.46	92.68
	9.07	1.47	89.46

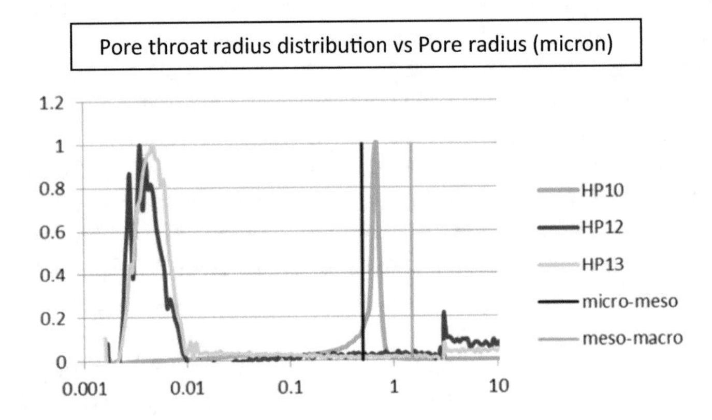

Fig. 10 Limestone Sopeña PTRD versus pore radius

determines the surface area of the sample and the distribution of pore diameters in the range from 0.3 to 300 nm. The referred area is called specific surface. Nitrogen is under critical temperature during the test, and solid surface is not in equilibrium, which will be reached as solid is saturated with gas. Test result is the physic gas adsorption until to achieve the thermodynamic equilibrium between the gas and the solid layer to define the specific surface (S_{BET}).

The interpretation of nitrogen adsorption-desorption isotherms is used to determine the pore size distribution (micropores and mesopores) of rock samples, as well as, obtain values of the total and external surface area of the material. Figure 11 shows adsorption-desorption isotherm from a nitrogen adsorption test carried out with Hontomín rock samples (Table 6).

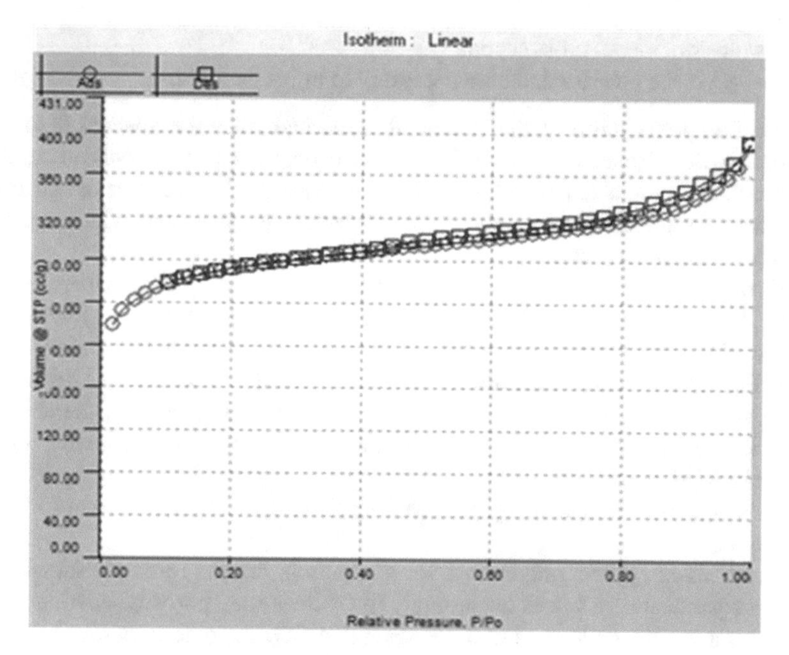

Fig. 11 Adsorption (red)-desorption (blue) isotherm to determine S_{BET}

Table 6 Specific surface of Hontomín seal-reservoir formations

Geological formation	S_{BET} (m^2/g)
Marly Liassic (cap-rock)	From 10 to 15
Limestone Sopeña (reservoir)	From 0.5 to 4
Dolomitic Sopeña (reservoir)	0.8

Accordingly data analyzed above, the main conclusion is Hontomín reservoir has very low open porosity values so that capillary pressure is very high, what suggests CO_2 trapping is unlikely in the porous matrix, at least during the early phase of injection.

The permeability of the rock is related to its ability to allow fluid flow through the interconnected pore system (primary porosity) or the existing or induced fractures (secondary porosity). The absolute, or intrinsic permeability of the formation, is usually calculated by the following formula [21]:

$$Absolute\ permeability = Q\mu I/(A(p2 - p1))$$

where

- **Q** is the injected flow rate
- **μ** is fluid viscosity
- **I** is the core logger height

- **A** is the core logger cross section
- **p2 – p1** is the pressure increase necessary to inject the fluid.

Effective permeability is the value that is reached when the injected fluid is in multiphase flow. Relative permeability is the ratio between the effective and absolute permeability, being able to reach values between 0 and 1. The relative permeability provides relevant information about the displacements of a fluid with respect to another during the injection.

Absolute permeability was determined for the study case by Klinkenberg Permeability Test [22]. Figure 12 shows the plot measured permeability (Kg) versus the inverse of average gas pressure (1/Pm), corresponding to Hontomín reservoir.

As Fig. 12 shows, the permeability increase is directly proportional to the inverse of the average gas pressure. In the case of using a liquid in the test, the absolute permeability is independent of the injection pressure. Permeability tests using liquids will be tackled in Sect. 4.2.

Table 7 shows Klinkenberg permeability ranges of Hontomín seal-reservoir formations.

Marly Liassic shows proper data for a cap-rock, but the reservoir shows very poor gas permeability values accordingly to the low open porosity existing in the tested samples, which lead us to think again that CO_2 trapping in porous matrix is unlikely. Therefore, liquid injection under reservoir condition is necessary to analyze

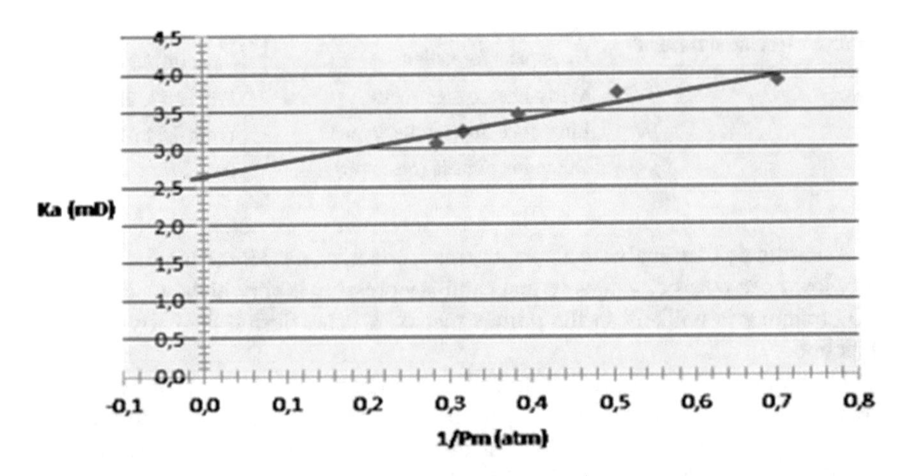

Fig. 12 Gas permeability versus inverse of average gas pressure for Hontomín reservoir sample

Table 7 Klinkenberg permeability values of Hontomín site

Geological formation	Kg (mD)
Marly Liassic (cap-rock)	0.001
Limestone Sopeña (reservoir)	From 0.001 to 0.5
Dolomitic Sopeña (reservoir)	From 0.001 to 0.006

hydrodynamic and mechanical effects both on porous matrix and fracture network, which may give additional clues on CO_2 migration in Hontomín reservoir.

4.1.3 Geomechanical Properties

Geomechanics provide valuable information to characterize the seal-reservoir pair for CO_2 storage, particularly, to define the stress-strain state and the changes induced by the injection in the rock massif. Geomechanical properties are defined by uniaxial and tensile strength, Young modulus, Poisson coefficient, cohesion and internal friction angle. These data, as others from laboratory works, are used to design field scale tests, build the geomechanical modeling [23] and define safe injection strategies.

Uniaxial Compression Test (UCT) is intended to define the stress-strain evolution and point load strength for an unconfined sample [24]. Rock specimen is usually cylindrical in shape subjected to a homogeneous load between its parallel bases until breakage occurs. Sample height must be between 2.5 and 3 times larger than its diameter. Vertical stress is the uniaxial compressive strength value at breaking point, and horizontal stress is zero as the sample is unconfined.

Loading speed was set in the range 10–15 MPa/min, accordingly the standard ASTM D2938, using two axial and one radial extensometers to measure the specimen deformation. Figure 13 shows vertical stress-strain graph and breaking point load for a sample from Hontomín cap-rock. Figure 14 shows the graph with vertical stress versus axial, lateral and volumetric strains.

Table 8 shows average values of UCT from samples of main Hontomín formations.

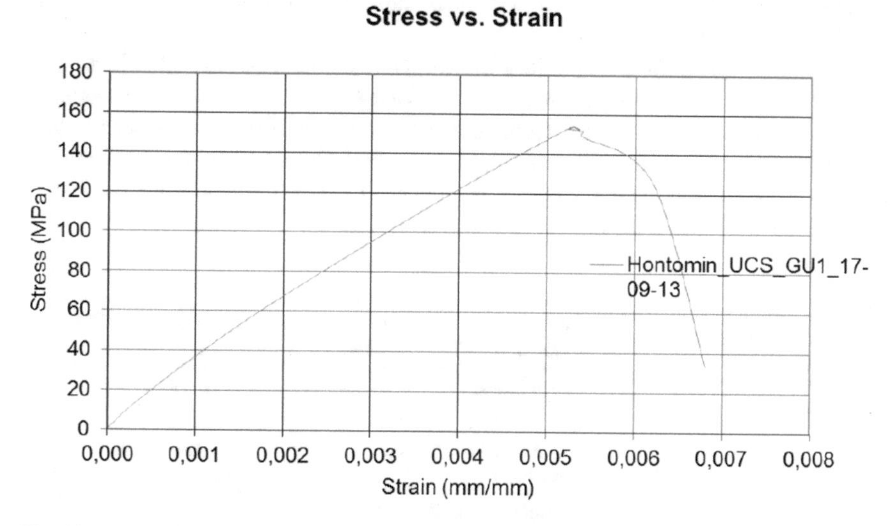

Fig. 13 Vertical stress versus vertical strain and breaking load point of cap-rock sample

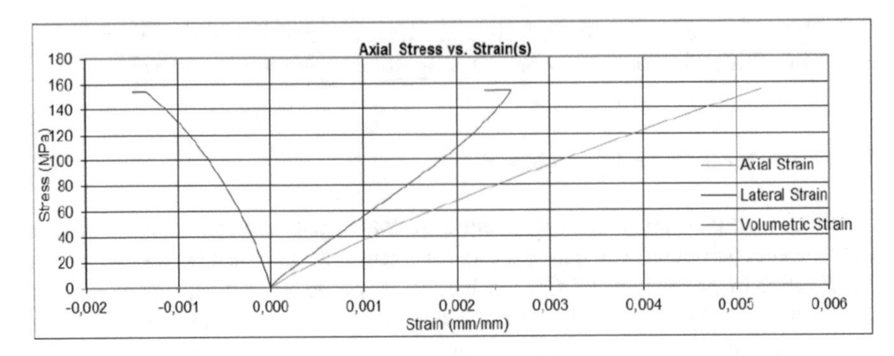

Fig. 14 Vertical stress versus axial, lateral and volumetric strain of cap-rock sample

Table 8 Uniaxial compressive strength of Hontomín formations

Geological formation	UCS (MPa)
Marly Liassic (cap-rock)	≥130
Limestone Sopeña (reservoir)	≥180
Dolomitic Sopeña (reservoir)	≥190

UCS values seem high for carbonated marl as Marly Liassic is and carbonates of Limestone and Dolomitic Sopeña, which probably have lower values in the outcrops, but for confined rocks at 1500 m depth as Hontomín seal-reservoir complex are quite usual.

Other parameters determined by UCS tests are Young Modulus and Poisson Coefficient. Table 9 shows average values of these parameters corresponding to Hontomín geological formations.

UCS and Young Modulus values of seal samples define Marly Liassic as high strength and medium stiffness rock massif, what suggests the associated tectonics did not produce a high fracture degree in the domed formation. Nevertheless, values corresponding to Sopeña reveal very high both strength and stiffness what suggests relevant fracture making in the reservoir produced by the dome setting-up (see Fig. 2 of Chapter "On-Site Hydraulic Characterization Tests").

Table 9 Young modulus and Poisson coefficient of Hontomín site

Geological formación	Young modulus E (GPa)	Poisson coefficient υ
Marly Liassic (cap-rock)	15–30	0.19–0.26
Limestone Sopeña (reservoir)	30–60	0.29–0.43
Dolomitic Sopeña (reservoir)	50–85	0.23–0.4

Table 10 Load ratio, confinement pressure and axial strength in TCT on Hontomín samples

Load ratio (MPa/min)	Confinement pressure (MPa)	Axial strength (MPa)
25	13	250–300
30	25	300–350
35	36	350–450

Table 11 Operating parameters of TCT carried out on a cap-rock sample

Maximum load (kN)	Axial stress (MPa)	Axial strain (mm/mm)	Lateral strain (mm/mm)	Volumetric strain (mm/mm)	Confinement pressure (N/mm^2)
276.896	219.358	0.007	−0.002	0.003	13.003

Triaxial Compression Test (TCT) is intended to determine the stress-strain state from a confined rock sample until the breaking point occurs [25]. Parameter like cohesion and internal friction angle may be determined using Mohr–Coulomb Criterion [26]. As occurs in UCT the rock specimen is usually cylindrical in shape subjected to a homogeneous load between its parallel bases and to a constant confinement pressure. A range of values was defined for the confinement of the samples tested, in order to determine the axial strength based on the applied load ratio. Table 10 shows the values used in TCT on Hontomín samples.

Table 11 shows results from TCT carried out on a Hontomín cap-rock sample.

Figures 15 and 16 show the graphs axial stress versus axial strain and axial stress versus axial, lateral and volumetric strains for the aforementioned triaxial compression test.

Results from UCTs and TCTs are coherent, generally showing high both uniaxial and triaxial strength values, increasing as the reservoir deepens. Young Modulus values increase in the reservoir as well, in relation to the existing ones of cap-rock.

4.2 *Injection Under Reservoir Condition*

CO_2 injection may impact on seal and reservoir changing the permeability and putting at risk the geological complex integrity. Therefore, injectivity and trapping mechanisms must be analyzed under reservoir condition at lab scale. Thus, data interpretation will be used to determine the most appropriate storage areas for CO_2 injection, check the suitability of the seal formation, design the tests to site hydraulic characterization and define efficient and safe injection strategies [4].

Biphasic flow of CO_2 and resident brine induce hydrodynamic, mechanical and geochemical effects both in porous matrix and fracture network of carbonates. These effects must be studied at lab scale using an equipment able to inject blends of CO_2 and brine under hydraulic and thermodynamic reservoir condition. The corresponding

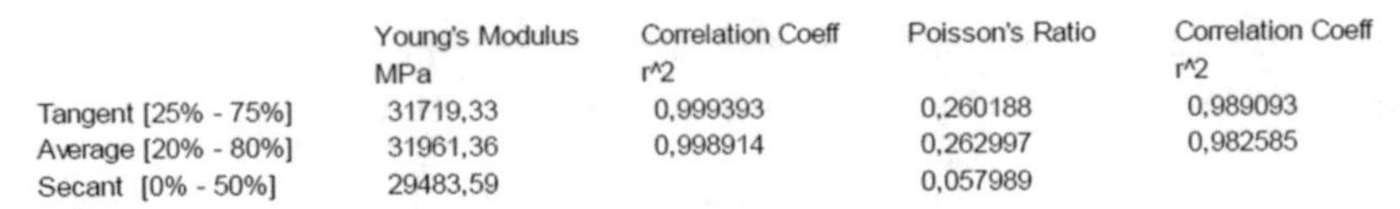

	Young's Modulus MPa	Correlation Coeff r^2	Poisson's Ratio	Correlation Coeff r^2
Tangent [25% - 75%]	31719,33	0,999393	0,260188	0,989093
Average [20% - 80%]	31961,36	0,998914	0,262997	0,982585
Secant [0% - 50%]	29483,59		0,057989	

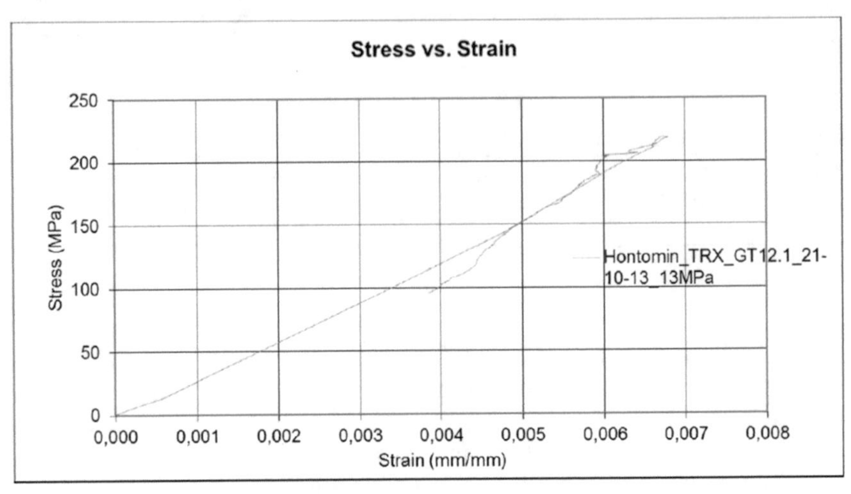

Fig. 15 Graph axial stress versus axial strain from TCT carried out on cap-rock sample

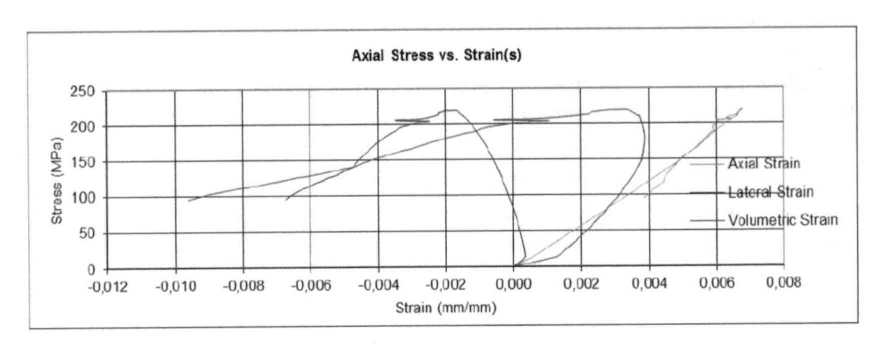

Fig. 16 Graph axial stress versus axial, lateral and volumetric strains from TCT carried out on cap-rock sample

results will be correlated with data from the laboratory works described in previous sections, and with those gained during the hydraulic characterization phase and early injection at field scale.

As mentioned in Sect. 4, ATAP equipment was used to carry out the co-injection of CO_2 and artificial brine both in dynamic and static operating modes, in order to analyze hydrodynamic, mechanical and geochemical effects induced on rock samples. In the next sections, tests performed and results obtained are described and analyzed.

4.2.1 Analysis of Hydrodynamic and Mechanical Effects

Accordingly ATAP protocol, CT Scanner and SEM are performed prior and subsequently to co-injections of CO_2 and brine in order to assess the induced changes both in the structure and composition of rock samples. CT Scanner and SEM data are used to identify the surfaces of each plug where fluid mixture comes in (injection surface) and out (production surface), paying special attention to the presence of open or filled fractures. Figure 17 shows CT Scanner image of Dolomitic Sopeña Plug in the injection surface prior to testing.

Moreover, a SEM test is performed on the injection surface of the plug to compare with changes after injection. Figure 18 shows SEM image prior to testing.

ATAP test parameters were the following:

- Confining pressure 140 bar
- Temperature 45 °C
- Rate artificial brine/CO_2 50% + 50%
- Flow rate 0.5 cm^3/min for both CO_2 and brine
- Injection pressure 75 bar.

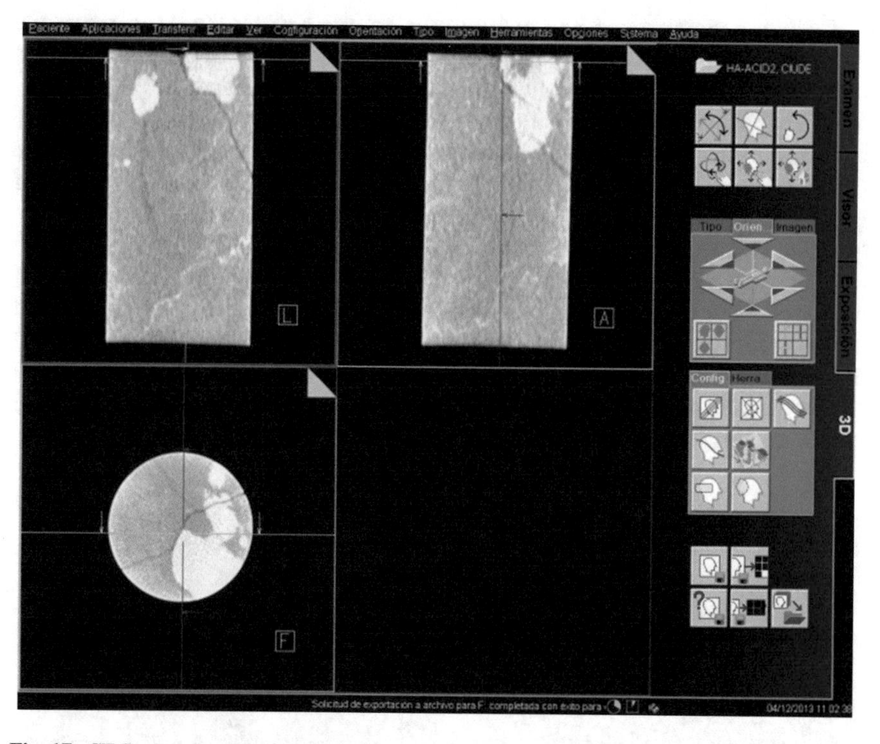

Fig. 17 CT Scanner image prior to testing

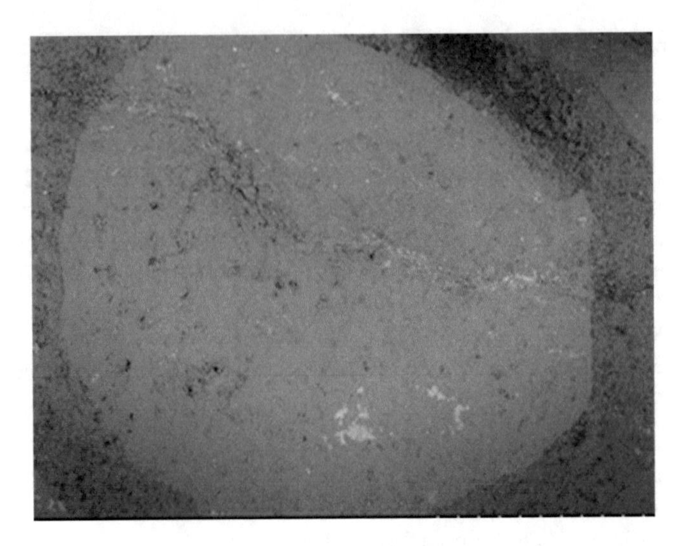

Fig. 18 SEM image prior to testing

Confining pressure and temperature were set accordingly the depth where the corresponding rock sample was extracted and data provided by well logging and other field tests performed during drilling (see Sect. 2.6 of Chapter "Light Drilling, Well Completion and Deep Monitoring"). Regarding the mixture composition of CO_2 and brine to inject in the plug, this depends on the goals pursued with the performance of the test, such as the determination of the relative permeability and the effects of geochemical reactivity on the rocks of the seal and reservoir (see Sect. 2.6 of Chapter "On-Site Hydraulic Characterization Tests").

There was a pressure increase of 1 bar at the beginning of co-injection of CO_2 and brine, both at the injection and production surfaces, what suggests the fluid transmissivity through the fractures, as gas permeability Kg was less than 0.005 mD in the sample. After pressure increase, fluid release occurred along the test what proved the fluid migration in the plug. Figure 19 shows the sample after co-injection of CO_2 and brine.

Fig. 19 Injection (left) and production (right) surfaces of the plug after co-injection

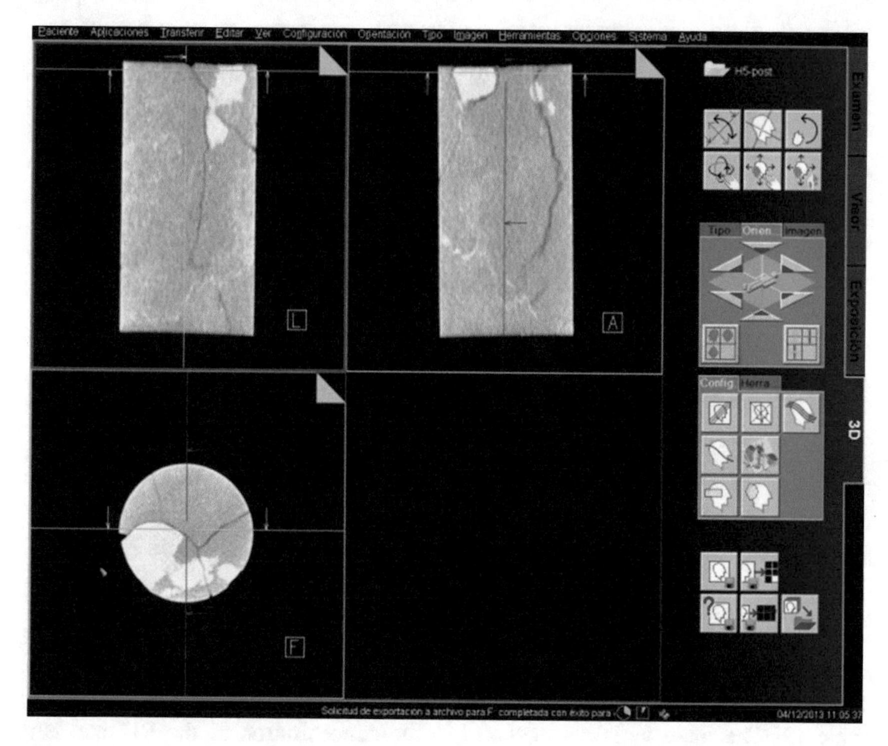

Fig. 20 CT Scanner image after testing

Figure 20 shows CT Scanner image in the injection surface after testing.

Figure 21 shows SEM image after testing.

Initially the plug showed fractures that were opened by the injection taking into account CT Scanner and SEM images. Injection pressure was 75 bar to maintain CO_2 in supercritical phase, what meant Leak off Test value was surpassed, since this parameter was in the range 60–70 bar for Sopeña Formation (see Sect. 5.3 of Chapter "On-Site Hydraulic Characterization Tests"). This fact could probably induce hydrodynamic and mechanical effects produced in the fracture network (i.e. dilatency and possible extension of fractures) [4].

On the other hand, geochemical reactivity was proved since the calcium sulfate layer on the production surface (see Fig. 19) has partially dissolved. This issue is addressed in the next section.

4.2.2 Analysis of Geochemical Effects

In addition to CT Scanner and SEM images taken before and after ATAP testing, effluent samples were also taken during the dynamic mode or injection and after the

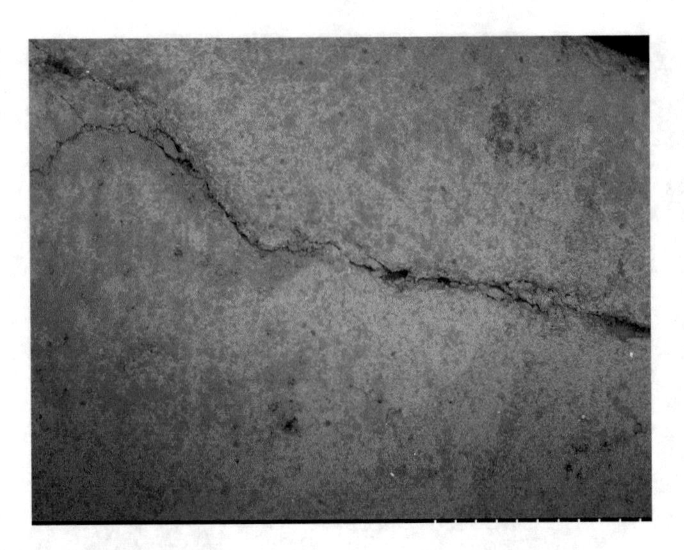

Fig. 21 SEM image after testing [4]

static mode during 20 day-trapping for chemical analysis. Mineral composition of rock sample was also determined using XRD, as mentioned above.

The effluents were analyzed by ion chromatography, showing the results in Table 12. The main variation occurred in the concentration of the Cl^- and Na^+ ions due to the use of a synthetic brine, injecting a constant concentration of the samples during the tests (30,000 ppm NaCl). On the other hand, the concentrations of SO_4^{2-}, Mg^{2+} and Br^- have increased considerably in the short term, and can be attributed to ionic migration [27].

Therefore, co-injection of CO_2 and brine under reservoir condition corroborated the permeability increase in the fracture network of samples due to hydromechanical effects and geochemical reactions which occurred in the short-term due to the interaction of the fluid mixture with the fillers of the fractures.

These results were used to design field scale hydraulic characterization tests at Hontomín, in order to overcome the low injectivity detected during well drilling (see Sect. 6.2 of Chapter "On-Site Hydraulic Characterization Tests").

Table 12 Chromatographic analysis of ATAP test effluent in ppm

Plug	ATAP phase	Cl^-	Br^-	SO_4^{2-}	Na^+	K^+	Mg^{2+}	Ca^{2+}
HA5	Pre-inyección	27.405	0	20	17.950	10	5	20
HA5	Post-inyección	33.390	10	35	21.885	10	22	15

5 Impacts of Impurities

The impurities existing in the captured CO_2 stream produce operating impacts during the injection, such as the injectivity decrease due to higher pressure is needed and less storage capacity since the gas density decreases [7]. Moreover, impurities usually produce chemical effects by dissolving the rocks that compound the seal-reservoir pair due to resident brine acidification, which puts at risk CO_2 geological trapping. Likewise, other chemical effects such the porosity decrease due to mineral precipitation, corrosion of transport and injection piping and well completion damage may occur. Therefore, the impacts of impurities must be studied both field and laboratory scale to assess the effects produced and design a contingency plan to mitigate them.

These studies were carried out at Hontomín site in IMPACTS Project [7, 28]. Regarding lab tests were intended to analyze the short-term chemical effects induced by the co-injection of CO_2-SO_2 in samples from the seal and reservoir. For additional information on physical effects produced by CO_2-N_2-O_2 injection at field scale, see the works and results detailed in de Dios et al. [7].

Mixtures of $CO_2 + SO_2$ (5% concentration) were injected under reservoir condition (confining pressure 150 bar/temperature 50 °C) in samples from Dolomitic Sopeña Formation (reservoir) saturated with artificial brine (NaCl 40,000 ppm) during 5 days, 8 h per day. Measurements of mass, porosity, Klinkenberg permeability, XRF and XRD were taken prior and subsequently the injection tests. Likewise, pre and post-injection CT Scanner and SEM images were captured, and ionic chromatography analysis performed on effluents after the injection. Tables 13 and 14 show the changes induced by the co-injection of CO_2-SO_2 in most representatives rock samples, both on physic properties and effluent ionic composition.

CO_2-SO_2 co-injection induced the porosity decrease probably due to the chemical reaction produced by the impurity, CO_2 and artificial brine with carbonates and sulfates that produced the mineral precipitation in the porous matrix. SO_2 in the mixture also produced the pH drop from slightly basic to strongly acidic, what suggests the short-term increase of some ions in the effluent, as $SO_4{}^{2-}$ and Mg^{2+}, due to the chemical reactivity between the mixture and the fracture fillings and rock matrix, taking into account the rock sample comes from Dolomitic Sopeña close to Keuper Formation (anhydrites).

Table 13 Changes of physic properties due to CO_2-SO_2 co-injection

Physic property	Pre-injection	Post-injection
Mass (g)	116.77	115.83
Porosity ø (%)	8.94	1.85

Table 14 Short-term evolution of effluent ionic composition

Anion concentration (ppm)					Cation concentration (ppm)					
Cycle	F^-	Cl^-	NO_3^-	SO_4^{2-}	Li^+	Na^+	K^+	Mg^{2+}	Ca^{2+}	pH
Brine	0.0	25,757	0.0	628.4	0.0	17,516	0.0	11.83	13.3	8.66
Brine-CO_2-SO_2	0.0	24,303	0.0	1961.9	0.0	16,013	4.5	144.1	13.1	1.75
Brine-CO_2-SO_2	0.0	23,829	0.0	4062.3	0.0	15,692	4.8	182.9	21.6	1.2
Brine-CO_2-SO_2	0.0	26,018	0.0	5700.4	1.2	17,144	1.4	194.8	13.4	0.98

6 Water Analysis

Water lab analysis involves both samples from shallow aquifers which usually supply fresh water to near populations, farms and industries, inter alia, and deep samples from the reservoir of the geological complex. Shallow water samples are analyzed in order to identify potential CO_2 leakages which can reach these aquifers close to the surface, and therefore, provide data to risk quantifying and apply mitigation tools. Resident brine analysis is intended to determine the chemical reactions produced by CO_2 injection in the pair seal-reservoir and quantify the trapping degree by solubility.

This section address the water analysis performed on resident brine samples from Hontomín reservoir. For additional information about Hontomín hydrogeological monitoring network and the shallow water analysis, see the works developed by de Dios et al. [13].

Resident brine sampling was performed using a U Tube [29], that is a permanent device installed in the injection well (HI) (see Fig. 2 of Chapter "Light Drilling, Well Completion and Deep Monitoring"), and according to the works carried out in the observation well (HA) by Flodim company using the removable device EZ BHS [30]. Correlation of data from both wells gives a realistic interpretation of CO_2 plume migration.

The samples are extracted in reservoir conditions (140 bar, 45 °C), transported and subsequently pressurized 2.5 times higher to maintain the sample integrity and measure the saturation pressure for that temperature. A flash evaporation is performed for each water sample in atmospheric condition (pressure $\leq$0.1 MPa and temperatures in the range 10–15 °C), to measure the ratio gas/liquid.

Gas phase is analyzed by ion chromatography. Table 15 shows the ion chromatography analysis of resident brine from the observation well (HA) at the early injection stage (2500 tons of CO_2 injected onsite).

Liquid phase was analyzed using API Ion Water Analysis Standard, to determine pH, temperature, conductivity and dissolved solids. Resident brine is slightly acidic at this stage with a pH close to 6.9 and dissolved solids in the range 38,540–39,230 mg/l.

Table 15 Ion chromatography of resident brine gas phase

Component	mol (%)
H_2	0
CO_2	7.041–43.028
N_2	10.374–52.059
CH_4	24.211–41.599
C_2H_6	3.015–5.265
C_3H_8	0.751–1.406
iC_4H_{10}	0.051–0.113
iC_5H_{12}	0.005–0.067
Benzene	0.0063–0.008
Toluene	0.023–0.076

Table 16 Ion chromatography of resident brine liquid phase

Components	Concentration (mg/l)
Cations	
Na^+	10,990–11,160
K^+	890–1010
Ca^{2+}	1810–1890
Mg^{2+}	700–780
Ba^{2+}	0.50–0.59
Sr^{2+}	36–38
Fe^{2+}	0.63–1.5
Anions	
Cl^-	20,210–20,770
SO_4^{2-}	3350–3400
HCO_3^-	250–310
Other components	
B^{3+}	16–17
Al^{3+}	<1
Si	9.9–12
P^{3+}	<0.3
Li^+	2.8–2.9

Ion chromatography analysis is also performed on liquid phase of the sample. Table 16 shows the range of most relevant ions of Hontomín resident brine.

Resident brine analysis shows low concentrations of CO_2 due to the early injection stage, with a salinity close to 40,000 ppm and a composition of Cl^- and Na^+ because of up to 14,000 m^3 of artificial brine was injected during the hydraulic characterization (see Chapter "On-Site Hydraulic Characterization Tests"), with relevant amounts of K^+, Ca^{2+}, Mg^{2+}, SO_4^{2-} and HCO_3^- by the carbonated nature of reservoir and the proximity of Keuper Formation (anhydrites).

7 Competency of Seal-Reservoir Pair

The seal-reservoir pair for CO_2 geological storage must assure the maximum amount of gas trapped in the long term. Therefore, the mineralogical and petrophysical composition of the rocks which compound the corresponding formations must be appropriate to accomplish these requirements along the project life (i.e. site characterization, injection, monitoring and abandonment).

de Dios et al. established in the study about the laboratory procedures for petrophysical characterization and monitoring of deep saline aquifers for CO_2 storage [13] the criteria to assess the competency of the cap-rock and seal. Tables 17 and 18

Table 17 Hontomín seal data and requirements

Properties	Requirements to be a seal	Hontomín seal
Most commom formations	Clayey and evaporitic	Carbonated marls
Injection pressure	70–90% LoT	Lower value
Open porosity (Φ)	<4%	From 2 to 4%
Klinkenberg permeability (Kg)	<0.1 mD	0.001 mD
Relative permeability CO_2 (Kr)	<0.01 for Sg >0.5	No data
Uniaxial strength (MPa)	>50	≥130
Young modulus (MPa)	20,000–40,000	15,000–30,000

Table 18 Hontomín reservoir data and requirements

Properties	Requirements to be a reservoir	Homtomín reservoir
Most common formations	Siliciclastic and carbonates	Carbonates
Injection pressure	>Capillary pressure	Not possible
Open porosity (Φ)	>10%	From 0.2 to 16%
Klinkenberg permeability (Kg)	>70 mD	From 0.001 to 0.5 mD
Relative permeability CO_2 (Kr)	>0.5 para Sg ≥0.5	No data
Uniaxial strength (MPa)	Not apply	≥180
Young modulus (MPa)	Not apply	30,000–85,000

show Hontomín site data of the seal and reservoir respectively, and the requirements that were set in the mentioned study.

Accordingly, Hontomín seal properties are usual to be the cap-rock for CO_2 storage. Nonetheless, the mineralogical composition must assure its integrity since the existing marls are carbonated up to 50% in some study cases, what means long-term lab tests must be developed [13] to assess whether chemical reactions induced by the resident brine and CO_2 on the carbonates could put at risk the seal integrity.

Hontomín reservoir properties are not aligned with the usual values for this type of geological complex for commercial purposes. It is a dual permeability medium with low effective porosity and hydraulic transmission dominated by the fracture network. This means it is really difficult to determine the site capacity due to modeling cannot give a clear idea about this parameter value. Moreover, the high anisotropy induced by fractures hinders a realistic prediction of plume migration.

8 Concluding Remarks

Laboratory scale works are part of the activities to carry out along the CO_2 storage project life. Data provide relevant information from the project viability decision making to the necessary conditions for a safe abandonment. This chapter address the lab works developed during the site characterization and early injection at Hontomín Technology Development Plant, being main concluding remarks the following:

- Hontomín main seal is composed by alternating sequence of marls, marly limestones, calcareous mudstones and shales with high degree of carbonation. Regarding the reservoir, it is composed at its upper part of mudstone/wackestone, with dolomite grainstone intercalations, and at the bottom by pure dolomitic grainstone and increasing anhydrite content because of the proximity of Keuper Formation.
- Marly Liassic samples show open porosity and gas permeability values of a seal formation, but those corresponding to Sopeña samples are not aligned to usual values of reservoir, since matrix porosity and permeability are really poor.
- Uniaxial compression strength and Young Modulus values define Marly Liassic as high strength and medium stiffness rock massif, what suggests tectonics did not produce a high fracture degree. Regarding the reservoir, values corresponding to Sopeña reveal very high both strength and stiffness what leads to think the relevant fracture making during the dome formation.
- Co-injection of CO_2 and artificial brine conducted in ATAP tests under reservoir condition produced hydrodynamic and mechanical effects in the fracture network (i.e. dilatency and possible extension of fractures), and geochemical reactions which occurred in the short-term due to the interaction of the fluid mixture with the fillers of the fractures. These effects generated permeability increase in the fracture network.
- SO_2 co-injected in the CO_2 stream induced the porosity decrease due to the chemical reaction produced by the impurity that generated the mineral precipitation in the porous matrix. Moreover, SO_2 in the mixture also produced the pH drop from slightly basic to strongly acidic, which increased the existence of some ions in the effluent because of fluid acidification.
- Resident brine analysis shows low concentrations of CO_2 due to the early injection stage at Hontomín site. Carbon dioxide was detected by ion chromatography after flash evaporation.
- Regarding results from lab works, Hontomín reservoir is a dual permeability medium with low effective porosity and hydraulic transmission dominated by the fracture network.

Some relevant lab works are still pending, such as the trapping tests or the resident brine and CO_2 impacts on the seal-reservoir pair in the long term, which will be performed in a more advanced phase.

Acknowledgements The experiences and results showed in this chapter form part of the projects "OXYCFB 300" and IMPACTS funded by the European Energy Programme for Recovery (EEPR), 7th Framework Programme (FP7) and the Spanish Government through Foundation Ciudad de la Energía-CIUDEN F.S.P. Authors acknowledge the role of the funding entities, project partners and collaborators without which the projects would not have been completed successfully.

This document reflects only the authors' view and that European Commission or Spanish Government are not liable for any use that may be made of the information contained therein.

Glossary

ATAP Alta Temperatura y Alta Presión
CT Scanner Computed tomography scanner
HA Observation well
HI Injection well
LOT Leak off test
MD Measured depth
PTRD Pore throat radius distribution
SEM Scanning electron microscopy
TCT Triaxial compression test
TDP Technology development plant
UCS Uniaxial compression strength
UCT Uniaxial compression test
XRD X-ray difraction
XRF X-ray fluorescence

References

1. Holt, R. M., Fjaer, E., Torsaeter, O., & Bakke, S. (1996). Petrophysical laboratory measurements for basin and reservoir evaluation. *Marine and Petroleum Geology, 13*(4), 383–391. Available online https://doi.org/10.1016/0264-8172(95)00091-7. Accessed January 28, 2020.
2. Dentith, M., Adams, C., & Bourne, B. (2018). The use of petrophysical data in mineral exploration: A perspective. *ASEG Extended Abstracts, Volume 2018, 2018—Issue 1: 1st Australasian Exploration Geoscience Conference—Exploration Innovation Integration.* Available online https://doi.org/10.1071/ASEG2018abW9_1F. Accessed January 28, 2020.
3. Noa, A. Z., & Shazly, T. (2014). Integration of well logging analysis with petrophysical laboratory measurements for Nukhul Formation at Lagia-8 well, Sinai, Egypt. *American Journal of Research Communication.* Available online https://doi.org/10.13140/RG.2.2.31300.12169. Accessed January 28, 2020.

4. de Dios, J. C., Delgado, M. A., Marín, J. A., Salvador, I., Álvarez, I., Martinez, C., & Ramos, A. (2017). Hydraulic characterization of fractured carbonates for CO_2 geological storage: Experiences and lessons learned in Hontomín Technology Development Plant. *International Journal of Greenhouse Gas Control, 58C*, 185–200.
5. Le Gallo, Y., & de Dios, J. C. (2018). Geological model of a storage complex for a CO_2 storage operation in a naturally-fractured carbonate formation. *Geosciences, 8*(9), 354–367. https://doi.org/10.3390/geosciences8090354
6. Porter, R. T. J., Fairweather, M., Pourkashanian, M., & Woolley, R. M. (2015). The range and level of impurities in CO_2 streams from different carbon capture sources. *International Journal of Greenhouse Gas Control, 36*, 161–174. Available online https://doi.org/10.1016/j.ijggc.2015.02.016. Accessed January 28, 2020.
7. de Dios, J. C., Delgado, M. A., Marín, J. A., Martinez, C., Ramos, A., Salvador, I., & Valle, L. (2016). Short-term effects of impurities in the CO_2 stream injected into fractured carbonates. *International Journal of Greenhouse Gas Control, 54P2*, 727–736. https://doi.org/10.1016/j.ijggc.2016.08.032
8. Eliassen, T., Richter, D., Crow, H., Ingraham, P., & Carter, T. (2008). *Optical and acoustic televiewer borehole logging. Improved oriented core logging techniques.* Available online https://www.marshall.edu/cegas/geohazards/2008pdf/Session2/04_geophys%20logging-TRB%20rev3%20(Diavik%20added).pdf. Accessed January 30, 2020.
9. Borgomano, J., Masse, J. P., Fenerci-Masse, M., & Fournier, F. (2013). Petrophysics of Lower Cretaceous platform carbonate outcrops in Provence (SE France): Implications for carbonate reservoir characterization. *Journal of Petroleum Geology, 36*(1). Available online https://doi.org/10.1111/jpg.12540. Accessed January 31, 2020.
10. Lombard, J. M., Azaroual, M., Pironon, J., Broseta, D., Egermann, P., Munier, G., & Mouronval, G. (2010). CO_2 injectivity in geological storages: An overview of program and results of the GeoCarbone-Injectivity Project. *Oil & Gas Science and Technology, 65*(65), 533–539. Available online https://doi.org/10.2516/ogst/2010013. Accessed February 4, 2020.
11. Ajayi, T., Salgado Gomes, J., & Bera, A. (2019). A review of CO_2 storage in geological formations emphasizing modeling, monitoring and capacity estimation approaches. *Springer Petroleum Science, 16*, 1028–1063. Available online https://link.springer.com/article/10.1007/s12182-019-0340-8. Accessed February 4, 2020.
12. Valle, L. (2012). ATAP design of a device for dynamic and static petrophysical studies of interaction between rock-brine-supercritical CO_2 in deep saline aquifers. In *GERG academic network event*, Brussels, Belgium, June 14–15, 2012.
13. de Dios, J. C., Álvarez, I., & Delgado, M. A. (2018). *Laboratory procedures for petrophysical characterization and control of CO_2geological storage in deep saline aquifers.* Spanish CO_2 Technology Platform (PTECO2) Publications. Available online https://www.pteco2.es/es/publicaciones/procedimientos-de-laboratorio-para-la-caracterizacion-petrofisica-y-el-control-de-almacenes-geologicos-de-co2-en-acuiferos-salinos-profundos
14. McPhee, C., Reed, J., & Zubizarreta, I. (2015). Routine core analysis (Chap. 5). In *Developments in petroleum science* (Vol. 64, pp. 181–268). Elsevier. Available online https://doi.org/10.1016/B978-0-444-63533-4.00005-6. Accessed February 5, 2020.
15. Malinverno, A. (2008). Core-log integration. In *Well Logging Principles and Applications G9947-Seminar in Marine Geophysics.* Spring 2008. Available online https://www.ldeo.columbia.edu/res/div/mgg/lodos/Education/Logging/slides/Core_log_integration.pdf. Accessed February 5, 2020.
16. Kovács, T. (2014). Characterization of Hontomín reservoir and seal formations. In *IV Spanish-French Symposium on CO_2Geological Storage*, May 13–14, 2014.
17. Salem, A., & Shedid, S. A. (2013). Variation of petrophysical properties due to carbon dioxide (CO_2) storage in carbonate reservoirs. *Journal of Petroleum and Gas Engineering, 4*(4), 91–102. Available online https://doi.org/10.5897/JPGE2013.0152. Accessed February 7, 2020.
18. de Oliveira, G., Ceia, M., Missagia, R., Archilha, N., Figueiredo, L., Santos, V., & Neto, I. (2016). Pore volume compressibilities of sandstones and carbonates from helium porosimetry measurements. *Journal of Petroleum Science and Engineering, 137*, 185–2016. Available online https://doi.org/10.1016/j.petrol.2015.11.022. Accessed February 10, 2020.

19. Mercury porosimetry. (2014). In *Tissue engineering* (2nd ed.). Science Direct. Available online https://www.sciencedirect.com/topics/engineering/mercury-porosimetry. Accessed February 10, 2020.
20. Kenneth, S. (2001). The use of nitrogen adsorption for the characterization of porous materials. *Colloids and Surfaces A: Physicochemical and Engineering Aspects 187–188*, 3–9. Available online https://www.eng.uc.edu/~beaucag/Classes/Nanopowders/GasAdsorptionReviews/ANotherReviewNotveryUseful.pdf. Accessed February 10, 2020.
21. Dake, L. P. (1998). *Fundamentals of reservoir engineering.*
22. Profice, S., Lasseux, D., Yannot, Y., & Hamon, G. (2012). Permeability, porosity and Klinkenberg coefficient determination on crushed porous media. *Society and Petrophysicists and Well-Log Analysts, 53*(6). Available online https://www.onepetro.org/journal-paper/SPWLA-2012-v53n6a5. Accessed February 11, 2020.
23. Olden, P., Pickup, G., Jin, M., Mackay, E., Hamilton, S., Somerville, J., & Todd, A. (2012). Use of rock mechanics laboratory data in geomechanical modelling to increase confidence in CO_2 geological storage. *International Journal of Greenhouse Gas Control, 11*, 304–315. Available online https://doi.org/10.1016/j.ijggc.2012.09.011. Accessed February 12, 2020.
24. Chau, K., & Wong, R. (1996). Uniaxial compressive strength and point load strength. *International Journal of Geomechanics, 33*(2), 183–188.
25. Kovari, K., Tisa, A., Einstein, H., & Franklin, J. (1983). Suggested methods for determining the strength of rock materials in triaxial compression: Revised version. *International Journal of Rock Mechanics and Mining Science and Geomechanics, 20*(6). Available online https://worldcat.org/issn/01489062. Accessed February 12, 2020.
26. Labhuz, J., & Zang, A. (2012). Mohr-Coulomb failure criterion. *Springer Rock Mechanics and Rock Engineering, 45*, 975–979. Available online https://link.springer.com/article/10.1007/s00603-012-0281-7. Accessed February 12, 2020.
27. Talman, S. (2015). Subsurface geochemical fate and effects of impurities contained in a CO_2 stream injected into a deep saline aquifer: What is known. *International Journal of Greenhouse Gas Control, 40*, 267–291. Available online https://www.sciencedirect.com/science/article/abs/pii/S1750583615001620. Accessed February 21, 2020.
28. IMPACTS Project. Available online https://www.sintef.no/projectweb/impacts/
29. Freifeld, B. (2009). The U-tube: A new paradigm for borehole fluid sampling. *Scientific Drilling, 8*. Available online https://www.sci-dril.net/8/41/2009/sd-8-41-2009.pdf. Accessed February 28, 2020.
30. FLODIM EZ BHS Technical Sheet. Available online https://www.flodim.fr/img/techsheets/EZ_BHS_2016.pdf. Accessed February 28, 2020.

On-Site Hydraulic Characterization Tests

J. Carlos de Dios, Carlos Martínez, Alberto Ramos, Juan A. Marín, and Jesús Artieda

Abstract Deep saline aquifers are target for carbon sequestration since these geological structures abound in many areas worldwide. Hydraulic characterization tests are focused on site feasibility assessment to inject CO_2 in an efficient and safely manner. For this, it is necessary to carry out both laboratory and field tests to determine hydraulic properties and operating parameters such as permeability and injectivity in the reservoir, and the trapping degree of the structural complex. CO_2 injection experiences usually come from projects conducted in aquifers composed by sandstones and similar rocks, unlike those carried out in carbonates that are quite limited. Sometimes carbonates are porous mediums, but in other cases, primary permeability is really poor being the fluid transmissivity mainly through the fracture network. Moreover, geochemical reactivity produced by the acidification of the mixture of CO_2 and resident brine plays a key role in these cases. This chapter address the innovative on-site hydraulic characterization tests conducted in the deep saline aquifer of Hontomín Technology Development Plant (Spain), which is composed of naturally fractured carbonates with low primary permeability. The impacts of artificial brine and CO_2 migration through the fracture network are described, analyzed and discussed, considering that produces hydrodynamic, mechanical and geochemical effects different from those caused by the injection in mediums with a high matrix permeability.

Keywords Carbon sequestration · Deep saline aquifers · On-site hydraulic characterization tests · Hontomín pilot · Naturally fractured carbonates · Poor primary permeability

J. C. de Dios (✉) · J. A. Marín
Foundation Ciudad de la Energía-CIUDEN F.S.P., Avenida del Presidente Rodríguez Zapatero, 24492 Cubillos del Sil, Spain
e-mail: jcdediosgonzalez@gmail.com; jc.dedios@ciuden.es

C. Martínez · A. Ramos
School of Mines and Energy, Technical University of Madrid, Calle de Rios Rosas 21, 28003 Madrid, Spain

J. Artieda
ARGONGRA, Paseo de San Francisco de Sales 38, 28003 Madrid, Spain

J. C. de Dios et al. (eds.), *CO_2 Injection in the Network of Carbonate Fractures*, Petroleum Engineering, https://doi.org/10.1007/978-3-030-62986-1_4

1 Introduction

CO_2 storage performed in deep saline aquifers composed of sandstones with high permeability in the rock matrix as Sleipner in Norway, In-Salah in Algeria and Decatur in USA represents the ideal conditions for establishing a commercial site accordingly the criteria defined in SACS and CO2STORAGE projects [1]. Other study cases in aquifers formed by carbonates with high primary permeability are AEP Mountaineer Project [2, 3], Michigan and Williston Basin CO_2-Enhanced Oil Recovery (EOR) projects [4, 5] in USA, IEAGHG Weyburn-Midale CO_2 project [6] in Canada and the Uthmaniyah CO_2-EOR demonstration project [7] in Saudi Arabia. However, there are not many sites in which the fluid transmissivity is mainly through the carbonate fractures as Hontomín case [8], a pilot where a total amount of 100,000 t of CO_2 is planned to inject.

As mentioned in Chapter "Light Drilling, Well Completion and Deep Monitoring", Hontomín Technology Development Plant is located in Northern Spain close to the city of Burgos. It is a structural dome with the reservoir and seal located at a depth from 900 (top of the dome) to 1832 m (flanks). Main seal is Marly Liassic and Pozazal formations and the reservoir is Sopeña formation [9]. A geological description of the site is given in Sect. 2 of Chapter "Light Drilling, Well Completion and Deep Monitoring".

Two wells were specifically drilled for the project, one for the injection (HI) and other for observation (HA), reaching the depth close to 1600 m. The distance between them is 50 m at surface and 180 m in the bottom hole. A detailed description of well completion and monitoring is given in Sect. 2.1 of Chapter "Light Drilling, Well Completion and Deep Monitoring". Four legacy wells are also located in the area (H1, H2, H3 and H4). Figure 1 shows the geological cartography including well location.

On-site hydraulic tests are part of site characterization process to assess the feasibility to inject and trap CO_2 in an efficient and safely manner accordingly the hydraulic property distribution of seal-reservoir pair [10]. Their main goal is to achieve field scale data to design CO_2 injection strategies appropriate to the geological complex characteristics. For this, it is necessary to study both the properties of CO_2 to inject (i.e. thermodynamics, flow rates and gas composition, inter alia) and petrophysical characteristics of seal-reservoir pair, paying special attention to hydrodynamic changes happened in the aquifer because of biphasic flow rate (CO_2/brine). It is particularly relevant for naturally fractured carbonates to study the impacts of CO_2 injection on geomechanical rock/fracture properties and the chemical reactivity induced by the acidification of resident brine [11].

Petrophysical laboratory works and hydraulic tests conducted while drilling at Hontomín revealed the low injectivity faced during preliminary tests [11]. How to manage the operations in this scenario was challenging, mainly due to the lack of experiences of CO_2 injection in carbonate fracture. 14,000 m^3 of artificial brine and 2300 t of CO_2 were injected on site during the period from May 2014 to December 2015, as part of the activities planned in the Project "OXYCFB 300, Compostilla"

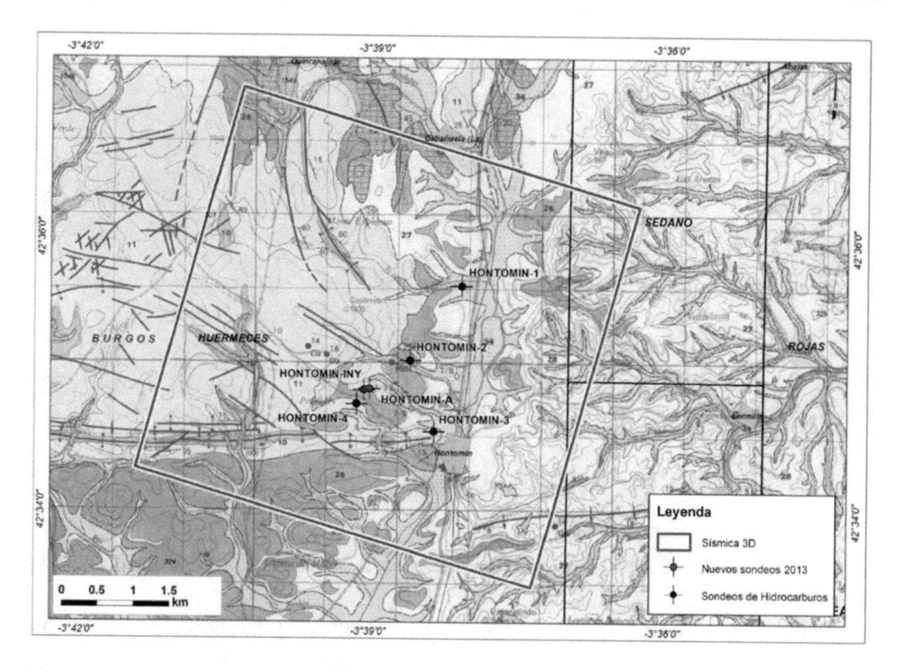

Fig. 1 Geological cartography of the study area

[8]. Innovative tests were performed to overcome the low injectivity faced during the hydraulic characterization phase, being needed the permeability enhancement in the fracture network that was carried out accordingly safe environmental criteria. Following sections address the methodology of tests performed, data interpretation and modeling used.

2 Hydraulic Characterization Goals

Final goal of hydraulic characterization is to design safe and efficient injection strategies regarding the captured CO_2 characteristics and the property distribution within the geological complex where gas will be permanently trapped. Nevertheless, previous steps are needed to reach this final goal.

First characterization works conducted at Hontomín as 2D-3D seismic campaign [12] and well logging [10], revealed the existence of two different faults and associated fractures groups which affect mainly the reservoir and overburden respectively, but there is no continuity in most of them through the cap rock and reservoir what ensures its integrity, according to the analysis performed by Le Gallo and de Dios [13]. Figure 2 shows main faults and fractures in the study area.

This fault arrangement proves the block compartmentalization of Hontomín site. For avoiding the interference in hydraulic transmission between the injection (HI)

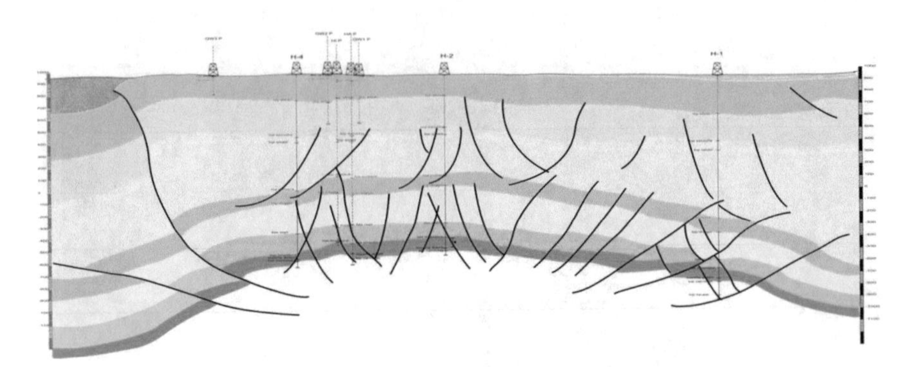

Fig. 2 Cross section of Hontomín geological complex

and observation well (HA), they were placed in the same geological block during the design phase, as Fig. 3 shows.

Firstly, it is necessary to check the fluid transmissivity in the block where the injection and observation wells are placed, and subsequently, to repeat this checking for the rest of wells used for CO_2 plume migration tracking. Therefore, first hydraulic tests will be performed between HI and HA to prove they are on the same geological block with hydraulic transmission. Next step is the hydraulic testing between the injection (HI) and legacy wells (H1, H2, H3 and H4) existing on site, to analyze the hydraulic property distribution in the study area.

Particularly, hydraulic properties of two main faults located at the South, known as Ubierna Fault, and East of Hontomín pilot must be studied. Both faults cross the storage complex from the reservoir to the overburden, what means they could be CO_2 leakage pathways. Obviously, the challenge is to prove if they are sealant or transmissive, determining both the horizontal and vertical permeability in the fault plane for each case. Figure 4 shows the geological modeling developed by Le Gallo and de Dios [13] including South and East faults.

It is also necessary to determine the permeability and transmissivity values along the open hole of each well, in order to identify the best areas for the injection and the expected capability of the geological formation (s) intended to be the cap-rock. Likewise, tests to determine the hydraulic fracture values of seal and reservoir formations are also needed to identify bottom hole pressures (BHP) corresponding to safe and efficient operating parameters.

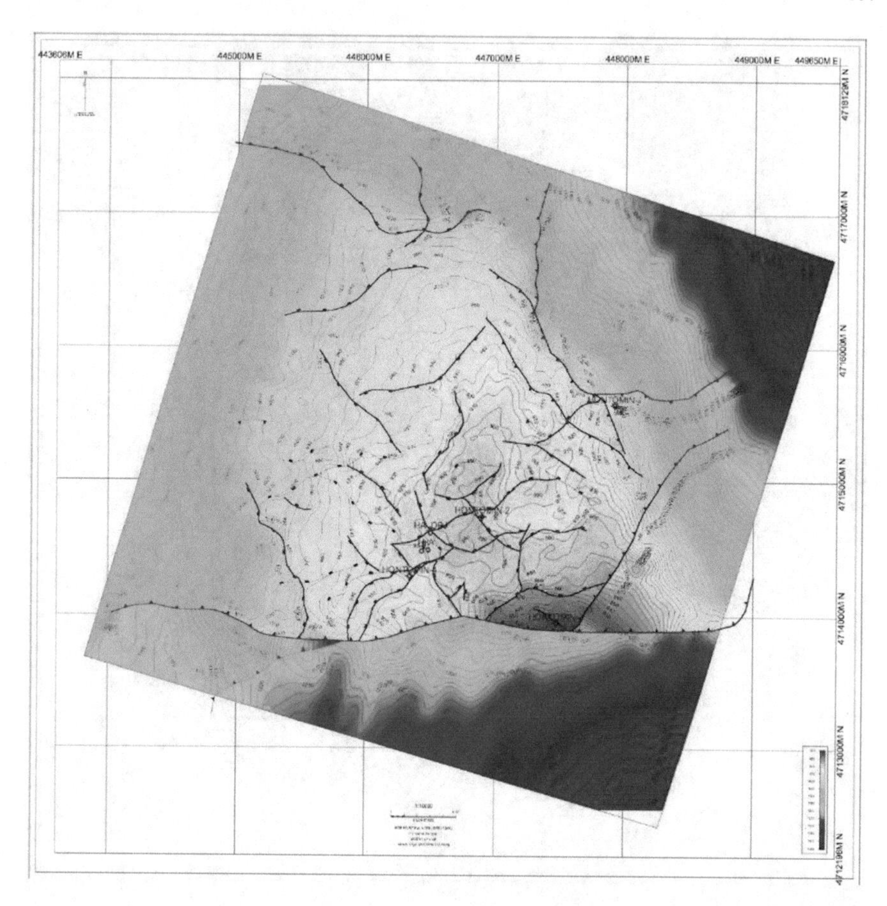

Fig. 3 Isochron layout of the top of Hontomín reservoir (Sopeña Formation) and the block where injection and observation wells (HI/HA) were located (red arrow)

3 Challenges to Overcome

The complexity of the structure of Hontomín seal-reservoir pair is in itself the main challenge to develop a fruitful hydraulic characterization of the site. Carbonate fractures bestow a high anisotropy level to CO_2 plume migration, what is enhanced in the study case due to the low matrix permeability of reservoir rock, what means the hydraulic transmissivity is dominated mainly by fractures [11]. Thus, the hydraulic property distribution is really complex to define at Hontomín, since in addition to the difficulty of modeling the fractured medium, the hydrodynamic and geomechanical effects produced by the CO_2 injection into the fracture network induce changes in the operating parameters. Moreover, chemical reactivity produced by the acidification of resident brine due to CO_2 injected in the aquifer must be considered. Mentioned effects will lead to changing injection conditions along the project life,

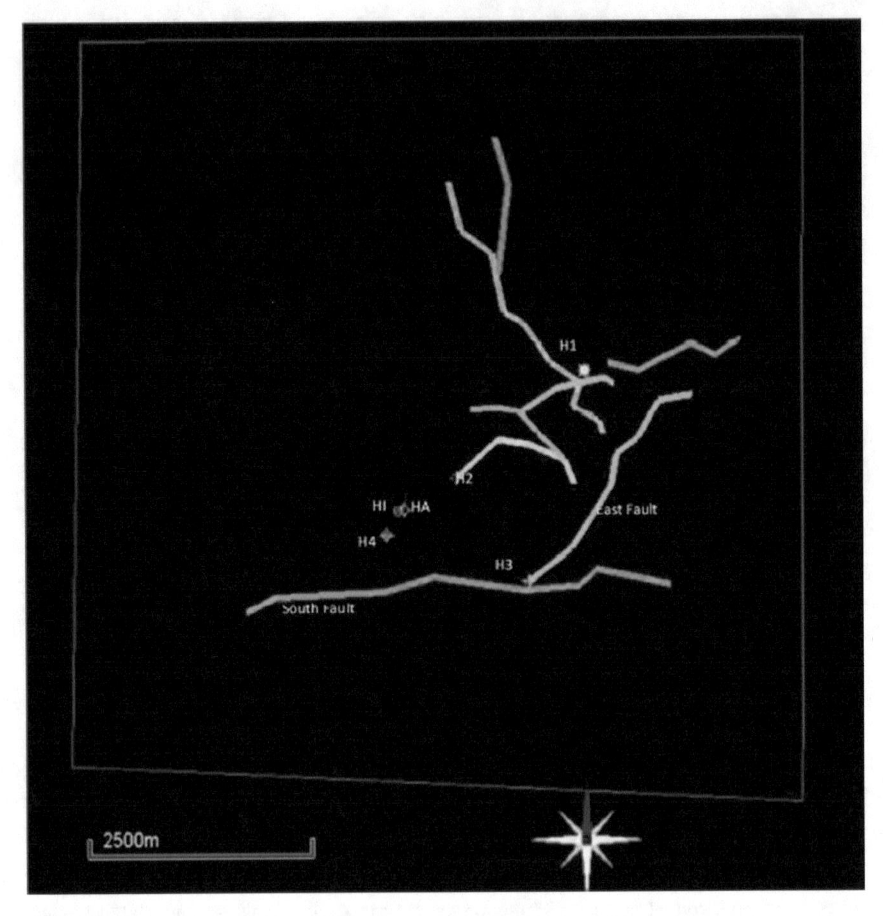

Fig. 4 Fault modeling of Hontomín study area. South fault (i.e. Ubierna Fault) in blue, and east fault in green

which researchers must bear in mind to gain conclusions from comparing actual and future scenarios.

On the other hand, hydraulic characterization of main faults (i.e. Ubierna and East fault) needs relevant amounts of brine and CO_2 to inject in order to achieve realistic data to interpret if they are sealant or transmissive, and in this last case, what would be the permeability values (i.e. horizontal and vertical) in each fault plane. Something similar occurs to check the hydraulic connectivity between the injection and legacy wells. Therefore, these objectives are likely not to be achieved during the characterization phase, being of main challenges to overcome during the early injection planned in ENOS Project, in which, a total amount of 10,000 t of CO_2 are planned to inject on site [14].

Undoubtedly, the complexity of Hontomín makes it an ideal site to investigate CO_2 storage in fractured carbonates [8, 11, 15, 16]. Its uniqueness forced researchers to develop innovative tests at both laboratory and well scale levels. It was also necessary to build geological [13] and fluid dynamic models that considered both the fractured nature of reservoir and the hydrodynamic and geochemical effects of the injection [17]. On the other hand, the role of geomechanical changes induced by the injection in the rock massif and fractures is a relevant activity to perform in future research works [18]. Results from Hontomín hydraulic characterization played a key role for granting the Storage Permit [19].

4 Inputs from Laboratory Works

As mentioned in Sect. 2.5 of Chapter "Light Drilling, Well Completion and Deep Monitoring", thirteen core samples were acquired from well drilling, ten from the observation well (HA) and three from the injection well (HI), of which seven correspond to the cap-rock (Marly Liassic and Pozazal Formations) and six to reservoir (four from Limestone and two from Dolomitic Sopeña Formations) (Table 2, Chapter "Light Drilling, Well Completion and Deep Monitoring").

Samples from cap-rock and reservoir formations were used to carry out laboratory tests to determine the injectivity and trapping capacity of the storage complex in the deep saline aquifer, accordingly the methods defined by de Dios et al. [20]. Chapter "Laboratory Scale Works" describes routine and specific petrophysical tests carried out on a laboratory scale and analyzes the results achieved.

Main routine tests performed during Hontomín characterization phase were the following (see Chapter "Laboratory Scale Works"):

- Core Logging
- Scanning Electrode Microscopy (SEM)
- Optical Microscopy (OM)
- X Ray Fluorescence (XRF)
- X Ray Diffraction (XRD)
- Density, porosity, pore size distribution and permeability
- Geomechanics (e.g. uniaxial strength, Young modulus, Poisson coefficient, inter alia parameters).

On the other hand, as mentioned above, CO_2 injection impacts on both cap-rock and reservoir samples, because it produces hydrodynamic and geochemical effects that may affect the seal integrity and induce permeability changes in the reservoir. For that, ATAP test [21, 22] was used to analyze the effects of injections of CO_2 and artificial brine under reservoir condition (Pressure = 140 bar and Temperature = 45 °C) on carbonate samples. More detailed description and explanation on ATAP test is given in Chapter "Laboratory Scale Works".

Before and after the text, samples were tested by scanning computer tomography (CT Scanner), 360° Photography and Scanning Electrode Microscopy (SEM). Moreover, when test is finalized the effluent (CO_2 + brine blends) was chemically analyzed. Thus, the interpretation of laboratory results will be used to determine the most appropriate storage areas for CO_2 injection, contrast the suitability of seal formation, to develop tests to hydraulically characterize the seal-reservoir pair and to design efficient and safe injection strategies.

Table 1 shows the results of petrophysical tests from rock samples of Hontomín seal and reservoir. The laboratory works performed were Mercury Porosimetry, Specific Surface BET and Klinkenberg Permeability.

Porosity is low in all cases, with pore diameters between 1 and 10 μm, typical of seal formations or reservoirs with low matrix porosity. Hontomín formations have low efficient porosity and high values of capillarity pressure, and accordingly, low gas permeability values. Main conclusion from these results is that Marly Liassic shows appropriate properties to be the seal, but further studies are necessary for checking whether Sopeña could be the reservoir.

Figure 5 shows SEM and CT Scanner images of a rock sample from Sopeña Formation, previously to carry out the CO_2 + brine injection under reservoir conditions by ATAP test.

The sample is representative of most of Sopeña Formation speciments, with low porosity (2.4%), abundant microporosity above 80% and virtually no mesoporosity,

Table 1 Petrophysical properties of Hontomín seal-reservoir formations

Formation	φ (%)	S_{BET} (m^2/g)	Kg (mD)
Marly Liassic (seal)	From 2 to 4	From 10 to 15	0.001
Limestone Sopeña (reservoir)	From 0.2 to 16	From 0.5 to 4	From 0.001 to 0.5
Dolomitic Sopeña (reservoir)	From 0.5 to 12	0.8	From 0.001 to 0.6

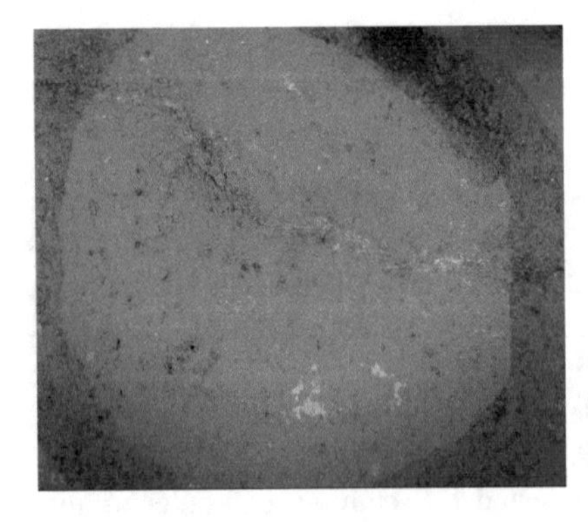

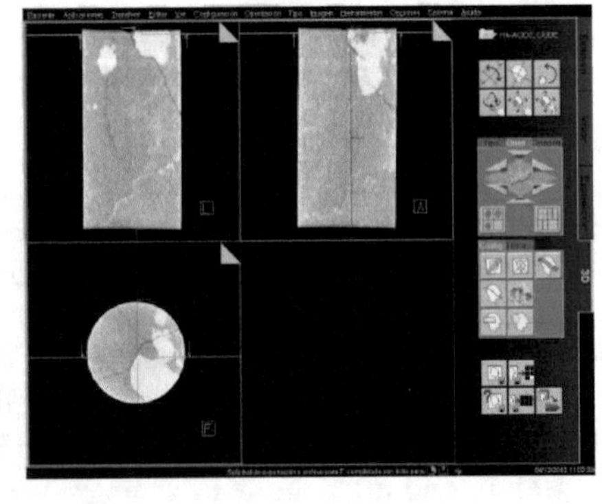

Fig. 5 SEM (left) and CT Scanner (right) images of Sopeña rock sample previously the injection

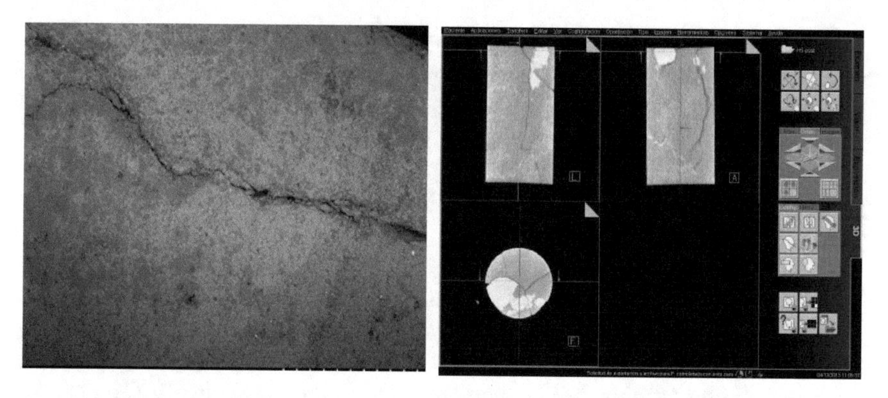

Fig. 6 Post injection SEM (left) and CT Scanner (right) images of Sopeña rock sample

and also low gas permeability (0.01 mD). The presence of fractures with different fillings is also a trend in Sopeña samples.

ATAP injectivity test was carried out firstly saturating the rock sample with artificial brine, with an equivalent salinity of 30,000 ppm NaCl, and subsequently injecting a blend CO_2 + brine (50% + 50%) with a pressure of 75 bar and temperature 45 °C to assure CO_2 supercritical phase. The sample was kept in the core logger for twenty days under a confinement pressure of 140 bar and temperature of 45 °C, according to Hontomín reservoir conditions. After this period, CO_2 + brine blend was released and the rock sample analyzed as did previously the injection. Figure 6 shows post-injection SEM and CT Scanner images.

Table 2 shows the results of effluent chromatography analysis once ATAP test finished.

Figure 6 shows changes in the fracture network, with open fractures that previously were joined or filled. Regarding the results from chromatography analysis, some ions increase in effluent composition. In some cases, such as Cl^- and Na^+, this effect may be caused by the sample saturation in brine, but in others, as SO_4^{2+} and Mg^{2+}, it is due to the acidification of the blend CO_2 + brine that produces the chemical reactivity with the rock, probably mainly within fracture fillings.

Therefore, the injection of CO_2 + brine under pressure (i.e. 75 bar) close to Leak off Test value could produce geomechanical changes mainly in fractures, inducing dilatency effects or extension of new fractures, enhancing the secondary permeability. Moreover, this fact was supported by the geochemical effects produced by the acidified brine on fracture fillings, such as the ion migration or dissolution [11].

Table 2 Results of effluent chromatography analysis

Phase	Cl^-	Br^-	SO_4^{2-}	Na^+	K^+	Mg^{2+}	Ca^{2+}
Pre-injection	27.405	0	20	17.950	10	5	20
Post-injection	33.390	10	35	21.885	10	22	15

Results from laboratory works allowed the design of well-scale hydraulic characterization tests suitable for a medium with low permeability, hydraulic transmissivity dominated by fractures and geochemical reactivity, as it is Hontomín carbonates case.

5 Tests While Drilling

As mentioned above, two wells were specifically drilled for OXYCFB 300 Project [8] at Hontomín, one for the injection (HI) and other for observation (HA), reaching the depth close to 1600 m below Dolomitic Sopeña Formation (see Figs. 1 and 2 of Chapter "Light Drilling, Well Completion and Deep Monitoring"). The distance between them is 50 m at surface and 180 m in the bottom hole. Figure 7 shows a panoramic view of Hontomín Technology Development Plant with well location and the injection and water conditioning facilities.

As described in Sect. 2 of Chapter "Light Drilling, Well Completion and Deep Monitoring", the wells were completed with the following deep monitoring devices (see Fig. 2 of Chapter "Light Drilling, Well Completion and Deep Monitoring"):

HI well

- 2 P/T sensors located at the open hole
- 1 Distributed Temperature Sensing System (DTS) anchored along the tubing
- 1 Distributed Acoustic Sensing System (DAS) anchored along the tubing
- ERT electrodes at the open hole
- 1 U-tube device for fluid sampling from bottom hole.

HA well

- 4 pressure/temperature (P/T) sensors in the seal and reservoir formations
- 28 ERT electrodes installed in the seal and reservoir formations.

Figure 8 shows the schemes of both wells with the completion and monitoring devices installed. A more detailed description of Hontomín well completion and monitoring is given in Sect. 2.1 of Chapter "Light Drilling, Well Completion and Deep Monitoring".

Fig. 7 Panoramic view of Hontomín Technology Development Plant

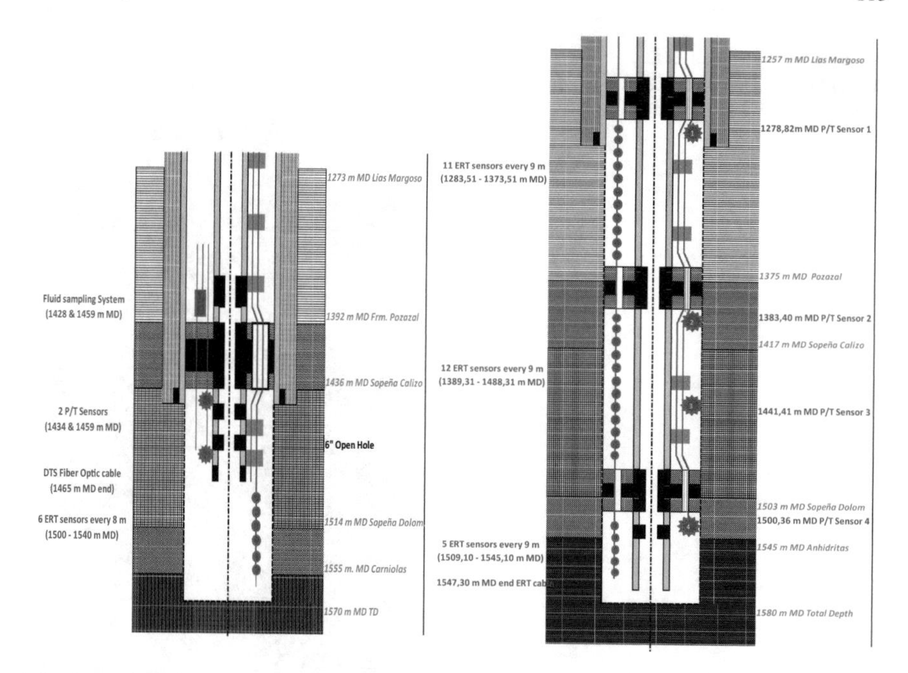

Fig. 8 Monitoring devices installed in the injection (left) and observation (right) wells

To achieve the goals described in Sect. 2 according to the results of laboratory works analyzed previously, the following on-site hydraulic tests were designed and performed during Hontomín well drilling [11]:

- Permeability test at field scale (PTFS).
- Connectivity test inter wells (CTIW).
- Leak off test (LOT).

5.1 *Permeability Test*

PTFS tests are commonly conducted during the last phase of well completion and provide information about the permeability in different parts of the open hole, in order to determine the reservoir area where the better CO_2 injection conditions are located. The equipment named Heavy Duty Double Packer system (HDDP) (Fig. 9) manufactured by the company SOLEXPERT [23] was used to carry out PTFS tests at Hontomín.

HDDP is formed by a threaded linkage string (tubing), two inflatable packers (top and bottom packers) and an injection nozzle placed between them. Brine is injected through the tubing by external pumps to test the permeability and transmissivity in the well area existing between both packers are placed.

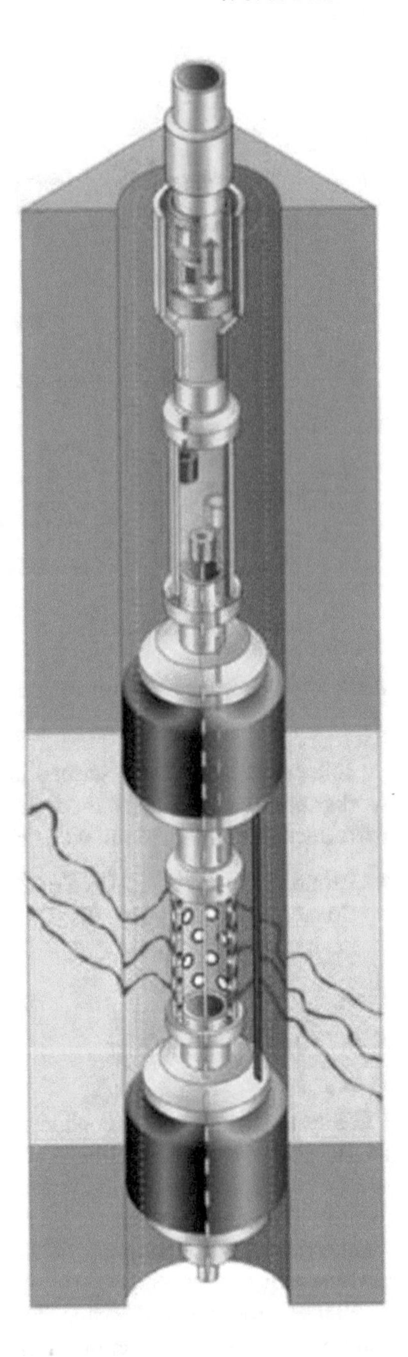

Fig. 9 Heavy Duty Double Packer system (Courtesy of SOLEXPERT)

Table 3 Permeability test at field scale (PTFS) conducted in HA well

Test number	Depth (m)	Injection period (h)	Fall off period (h)
1	1530–1580	8 h in total (injection and fall off)	
2	1501–1529.8	11.5	11.5
3	1472–1500.8	5	13
4	1439.1–1467.9	9	12
5	1414.2–1580	8.5	24

Table 3 shows HA well areas tested accordingly the depth where the top packer was anchored and the length set between top and bottom packers. A total of five tests were conducted on site to determine both permeability and transmissivity from 1414 m MD in the bottom of Pozazal Formation to 1580 m MD in the Anhydrites (see Figs. 1 and 2 of Chapter "Light Drilling, Well Completion and Deep Monitoring"). Thereby, the total length of the open hole where the reservoir (Limestone and Dolomitic Sopeña Formation) is located was tested. Artificial brine (salinity 30,000 ppm NaCl) was injected in periods of time showed in Table 3, measuring the pressure and temperature increase during each one, followed by the shut-it phase in fall-off periods to determine the recovery of both parameters.

Results from tests 1, 2, 3 and 4 (Table 3) were not promising to identify an appropriate area to efficient injection. Therefore, test 5 was conducted anchoring the top packer at the seal bottom holding the down packer deflected for allowing an injection area in the total depth range (1414–1580 m MD). This final test was carried out in pulses, first one using N_2 injection (35 bar, 30 s) followed by 90 min fall off to check the compressibility of fluid column. Afterwards, brine was injected under pressure of 40 bar during 60 min followed by 90 min fall off. Brine was injected in 3rd pulse with 12.5 bar during 4 h followed by 13 h fall off. In 4th pulse the brine was injected with 40 bar during 1 h followed by 3 h fall off. And finally, brine was injected in 5th pulse with 12.5 bar during 1 h. This final pulse was intended to check if injectivity increases after a higher energy pulse as 4th was. A slight increase in injectivity (17%) took place in the final test. Figure 10 shows the evolution of operating parameters during 5th test.

Table 4 shows the pressure measured during the injection in periods between 5 and 11.5 h depending on: the formation pressure and corresponding fluid densities, the depth where the gauge was set, the relation between pressure and gauge depth and the temperature in each interval. Data correspond with the injection of brine in the different parts of the observation well (HA) listed in Table 3. The interpretation was performed with a radial composite model (Saphir™, Kappa) and results are fluid transmissivity (Kh), permeability (K) and skin factor (S) [11]. Section 7 address a detailed analysis of modeling developed for PTFS tests and others conducted during Hontomín hydraulic characterization.

Permeability and transmissivity values are higher using the total well length from seal bottom to Anhydrites, and this was the reason to complete HI bottom in open hole (see Fig. 8). Although these parameters have low values, what was revealed is

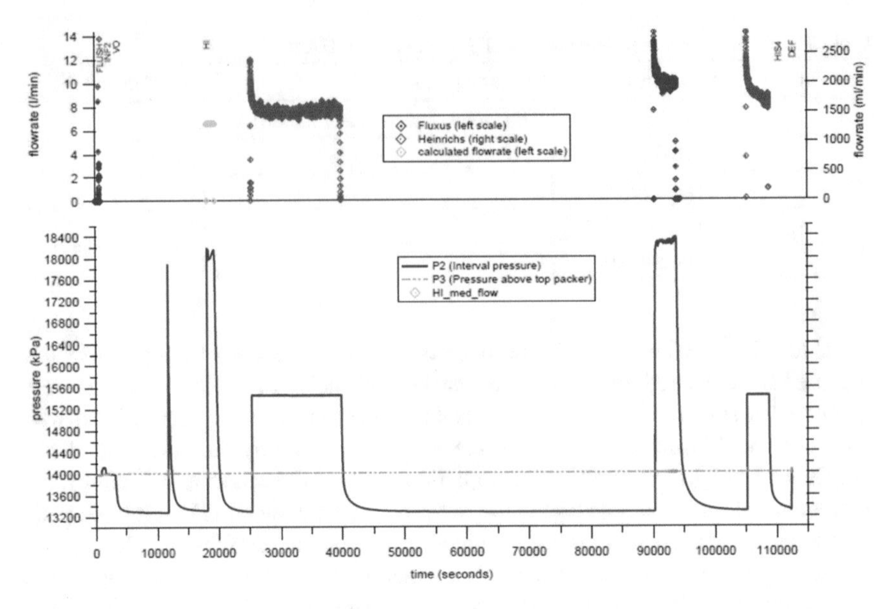

Fig. 10 Bottom hole pressure (BHP) (down) and flow-rate (up) values during 5th PTFS test

BHP increase in test 4th induced a higher injectivity. This fact could be related to hydrodynamic and geomechanical effects produced by the injection in the fractures, as the laboratory works analyzed above shown.

Hereinafter to prove the hydraulic connectivity between the injection and observation wells and to determine the hydraulic fracture value (leak off test) for seal and reservoir formations were necessary.

5.2 *Connectivity Test Inter-Wells*

As mentioned in Sect. 2, accordingly the prognosis provided by the geological modeling it was intended to place the injection and observation wells in the same geological block (see Fig. 3). CTIW tests were used to prove whether both wells were right located as planned by checking of hydraulic transmission existence between them.

Through these tests the pressure correlation between the injection and the observation well (HI and HA) is determined. For this, PERFRAC 134 equipment [23] was used in the HA well to record pressure and temperature data, using an inflatable packer above these probes to close the bottom hole during the test. PERFRAC 140 [23] equipped with injection nozzle and an inflatable packer able to withstand a differential pressure of 300 bar was installed in the open hole of HI well to produce the pulses and their monitoring.

Table 4 PTFS results during brine injection in HA well

Depth (m)	Height of range (m)	Final pressure (bar)	Measure depth	Ratio pressure/depth (bar/m)	Temperature (°C)	Transmissivity (mD m)	Permeability (mD)	Skin factor
1530–1580	50	–	1530	–	46.5	–	–	–
1501–1529.8	25.8	141.2	1501	0.0941	45.8	0.18	0.0225	−1
1472–1500.8	28.8	138.6	1472	0.0942	44.8	1.43	0.286	0.4
1439.1–1467.9	28.8	135.4	1439.1	0.0941	43.9	0.95	0.063	9.8
1414.2–1580	165.8	132.5	1414.2	0.0938	43.3	24.23	0.866	1

Figure 11 shows the scheme of PERFRAC 140 installed in the open hole of the injection well (HI), including tubing 2 7/8″, an upper packer anchored in 1435.55 m MD and the extension formed by the memory gauge (P/T) placed in 1436.26 m MD and the mandrill containing downhole pressure gauge placed in 1438.31 m MD in the upper part of reservoir.

Observation well arrangement using PERFRAC 134 was quite similar using a cable to run the equipment along the well and for data transmission, installing the upper packer in 1278.66 m MD and the downhole pressure gauge placed in 1280.71 m MD.

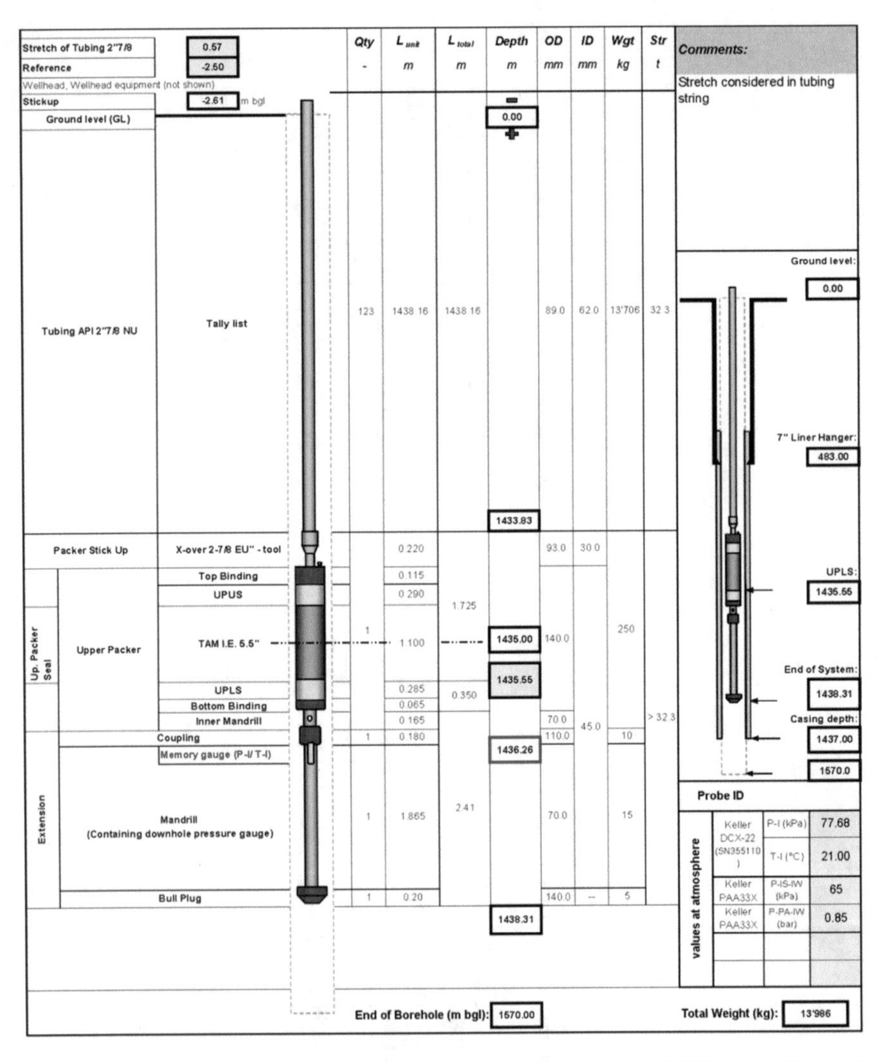

Fig. 11 PERFRAC 140 installation in the injection well (HI) to perform CTIW tests at Hontomín (Courtesy of CIUDEN)

Table 5 1st CTIW test conditions

Brine flow rate (l/min)	Injection period (min)	WHP initial (bar)	WHP final (bar)	Fall off (min)
2				
60	30	30	37	
80	30	43	44	
100	30	46	50	
120	60	53	56	60
120	60	34	58	60
120	60	38	60	20
120	10	52	57	660

2 CTIW tests were conducted for checking the hydraulic connectivity between the injection and observation wells using artificial brine (30,000 ppm NaCl). 1st test was carried out under the conditions detailed in Table 5.

Figure 12 shows the bottom hole pressure evolution in both wells according to the injected flow rates during the first test. Initial BHP evolution is clearly sinusoidal with peaks followed by pressure decreases what reveals the opening of fractures induced by brine injection. On the other hand, accordingly the values recorded in the injection and observation wells, there is a delay of 20 min between the injection pressure in HI

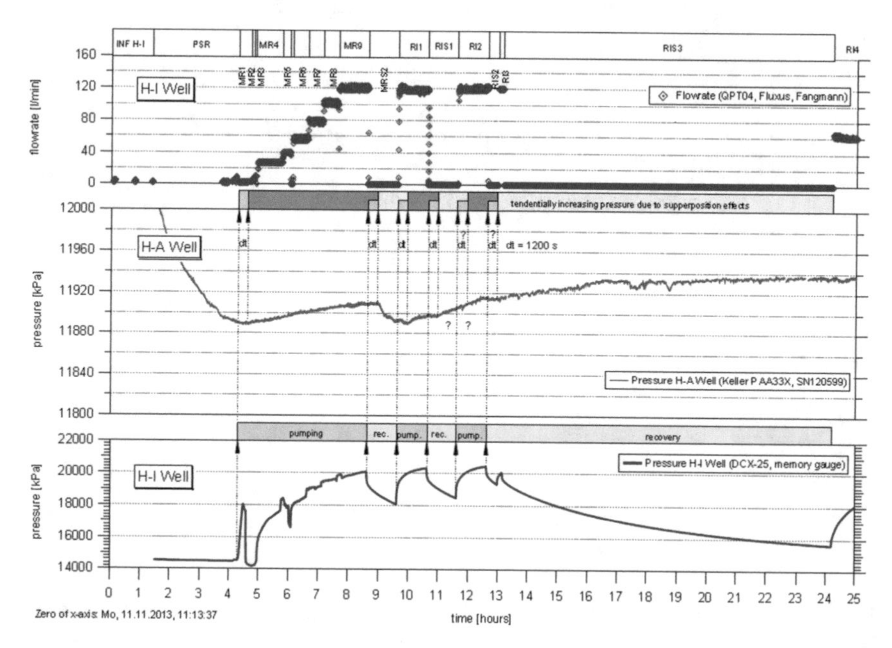

Fig. 12 Evolution of BHP in HI well (down), BHP in HA well (middle) and flow rate (up) during 1st CTIW test

well and the pressure values recorded in HA well. Taking into account the distance between both wells at the bottom hole is 180 m, this fast response to injection in the observation well lead us to think the hydraulic transmissivity is dominated by fractures [11].

2nd test was performed by 3 pulses of brine injection followed by fall off periods as shown in Table 6. After last fall off of 24 h the well head pressure decreases from 67 bar to the atmospheric pressure. While packer deflating in HI well a BHP decrease was detected in HA well, as Fig. 13 shows, what proves the hydraulic connectivity between both wells [11].

The location of both wells in the same geological block with hydraulic connectivity between them was proved regarding the results from CTIW tests analyzed above.

Table 6 2nd CTIW test conditions

Brine flow rate (l/min)	Injection period (min)	WHP initial (bar)	WHP final (bar)	Fall off (min)
60	60	13	38	120
120	60	19	53	120
180	120	30	67	1440

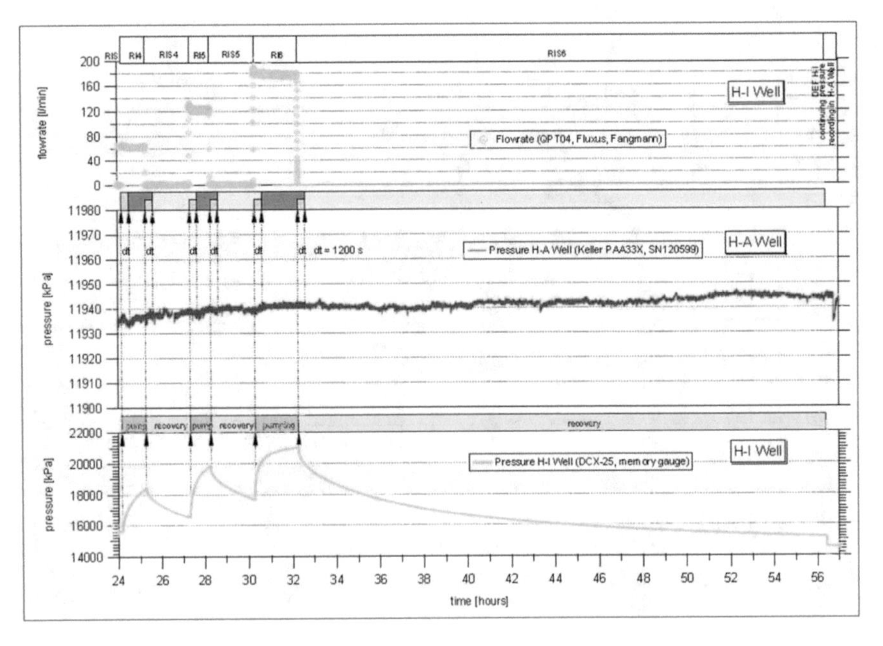

Fig. 13 Evolution of BHP in HI well (down), BHP in HA well (middle) and flow rate (up) during 2nd CTIW test

Similar tests should be developed between the injection well (HI) and the existing legacy wells (H1, H2, H3 and H4) according to the expected CO_2 plume migration. As mentioned before, these works must be carried out during several phases of pilot operation to inject the planned amount of 100 kt of CO_2.

5.3 Leak-Off Test

Leak-off test (LOT) determines the leak-off pressure (LOP) which is used to evaluate the minimum stress that produces the rock massif fracture in a borehole and to determine the in-situ stress tensor [24].

LOT is the usual test carried out during well drilling to determine the hydraulic fracture of rock formations along the boreholes where petrophysical characteristics have not been established yet. LOP values achieved from tests are used for purposes related to drilling (e.g. define mud density to avoid rock hydraulic fracturing) and also to design efficient and safe injection strategies.

In case of deep wells where reservoir condition of pressure and temperature are usually high, CO_2 injection is conducted in dense phase (i.e. supercritical or liquid) which produces the BHP increase necessary to open the fractures within the reservoir. Moreover, injection must be carried out in efficient and safe conditions, which means the cap-rock integrity needs to be assured. Reservoir LOP value gives information to inject CO_2 in an efficient manner, since in tight reservoir cases where fluid migration is dominated by fractures the injection pressure close to LOP produces new fractures which enhance permeability. But if the induced fractures in the reservoir reach the cap-rock formation they can produce leakage pathways [25]. Therefore, the knowledge of LOP values both cap-rock and reservoir is crucial to define a safe BHP range during the whole life of the project.

LOT procedure followed in Hontomín was as follows:

1. Casing/liner testing prior to drilling out the shoe
2. Drilling the shoe + cement
3. Mud circulating
4. Bit lift and installation of pump + lines
5. BOP closure
6. Start of pumping
7. WHP/Flow rate monitoring
8. Bleed off the pressure and establish the amount of mud injected.

LOT conditions performed in the injection well (HI) at 1437 m MD (Pozazal Fm bottom and Top of Sopeña Fm) are shown in Table 7. Figure 14 shows WHP monitoring graph of mentioned test.

Table 7 HI LOT 7″ liner shoe at 1437 m MD, 6″ open hole

Mud density (g/cm^3)	1.05
Depth (m MD)	1437.12
Hydrostatic pressure (bar)	148.03
LOP (bar)	55
Fracking BHP (bar)	203.03
Fracking gradient (bar/m)	0.14
Equivalent mud weight (g/cm^3)	1.44

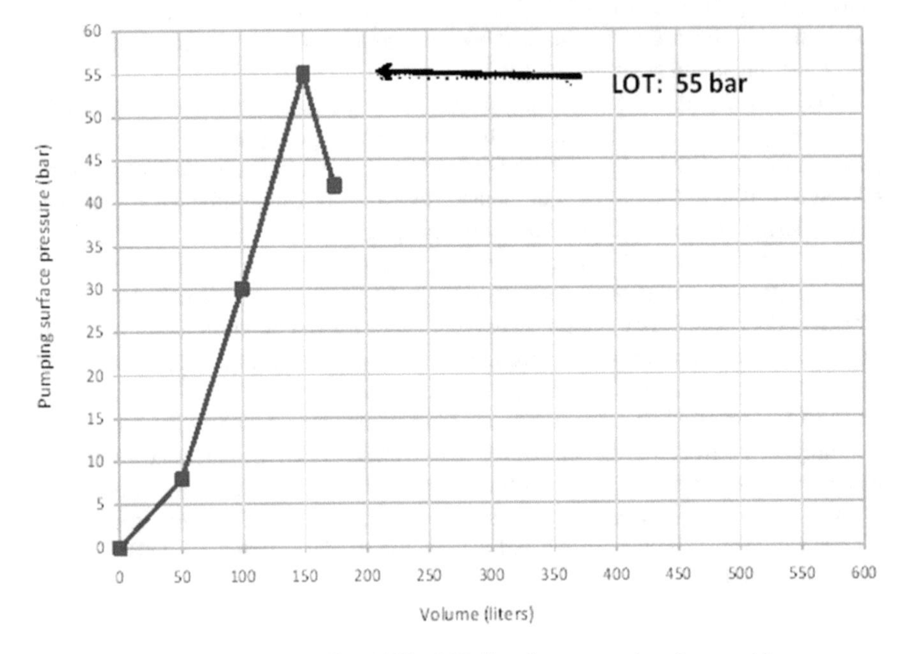

Fig. 14 LOT performed in HI well at 1437 m MD (interface cap-rock and reservoir)

Regarding LOTs conducted while HI/HA drilling, LOP average values for the cap-rock and reservoir of Hontomín are the following:

- Cap-rock LOP average values = 50–70 bar
- Reservoir LOP average values = 60–70 bar.

6 Test to Determine the Injectivity in the Reservoir

Results from tests while drilling proved the following:

- Both wells for injection and observation (HI/HA) are located in the same geological block with hydraulic transmission between them

- Hydraulic properties of reservoir are typical of double porosity medium, with very poor primary permeability and fluid transmissivity dominated by fractures
- Well head pressure (WHP) value of injected fluids above 70 bar exceeds LOT threshold, both in the reservoir and cap-rock, what means hydraulic fracturing may occur altering the hydraulic properties and putting at risk the site safety.

On the other hand, laboratory scale results lead us to think the injection of a blend of CO_2 in supercritical phase and brine may produce hydrodynamic, geomechanical and chemical changes in the fracture network, enhancing the secondary permeability in the reservoir.

According to mentioned above, in order to determine the injectivity range in the reservoir to design safe and efficient injection strategies, following field scale tests were designed and performed:

1. Brine injection
2. Co-injection of brine and CO_2.

14,000 m^3 of artificial brine and 2300 t of CO_2 were injected at Hontomín during the period from May 2014 to December 2015, as part of the activities planned in the Project "OXYCFB 300, Compostilla" [8].

6.1 *Brine Injection*

Unlike the usual hydraulic field tests which are carried out in production mode [26], what means the resident brine is pumped out from the reservoir, the tests performed at Hontomín were in injection mode. Tests while drilling had already given clues about the hydraulic properties of the naturally fractured reservoir with low primary permeability. Obviously, the first reason to use injection tests is because of the inefficiency of preliminary production tests conducted during the drilling phase, which showed low flow rates and high recovery periods in the open hole of both wells (HI/HA). Therefore, it was decided to carry out tests injecting artificial brine, as used in the tests while drilling (30,000 ppm NaCl), as a first step previously the CO_2 injection in order to achieve the following goals:

- Identify the pressure/temperature allocation in the seal-reservoir complex accordingly the flow rates injected on site
- Analyze the geomechanical impacts of injections in the rock massifs and fracture network
- Define the pressure range to establish the injectivity threshold in a dual porosity medium
- Analyze the effect of injection pressure close to LOT value, particularly the extension of new fractures and the seismicity events
- Establish a preliminary injectivity abacus.

A core goal of reservoir hydraulic characterization is to set the top and lower limits of the injectivity range. Top limit is determined by the parameter values which correspond to the admissible seismic threshold and the fracturing level of the rock. The lowest pressure value corresponding to both effects defines the top limit. The lower limit is defined by the pressure-flow rate binomial corresponding to the injectivity threshold of reservoir that starts the fracture opening, which was defined by the ATAP test in the study case (see Chapter "Laboratory Scale Works"). Once the corresponding value setting has been adjusted accordingly CO_2 thermodynamic properties, the injection will be conducted during the pilot operation between both limits, correlating pressure values at the well head and bottom hole with the corresponding flow rates as established in the injectivity abacus. A matter to consider is the injectivity abacus is dynamic, showing operating parameters for particular scenarios which may vary along the project life.

Brine injection tests give a preliminary idea about the reservoir hydraulic behavior, taking into account the differences with CO_2 thermodynamic properties. The parameters to monitor during the tests are the following:

- Well head pressure (WHP)
- Well head temperature (WHT)
- Bottom hole pressure (BHP)
- Bottom hole temperature (BHT)
- Temperature along the injection tubing
- Brine flow rate.

To achieve the aforementioned goals two types of field scale tests under industrial conditions were designed:

- Brine injection tests in pressure control mode
- Brine injection tests in flow rate control mode.

Tests in pressure control mode are conducted holding WHP constant and analyzing how flow rate evolution is for this scenario. On the other hand, tests in flow rate control mode are conducted holding the flow rate constant and analyzing how WHP and BHP evolutions are.

BHP monitoring is crucial to design safe CO_2 injection strategies, using methodologies based on the existing well injectivity and using techniques which analyze bottom hole pressure evolution [3]. Therefore, injectivity threshold (WHP_i/Q_i) defined by ATAP tests as mentioned above, the rock fracturing threshold (WHP_f and Q_f) defined by LOT test and the seismicity effects produced by the injection, will be used as limits of the injectivity range during the implementation of field tests. Moreover, injection pressure (WHP_i) must be ranged between 70 and 80 bar at 10–20 °C of temperature in the well head to assure CO_2 injection in liquid phase, as required in OXYCFB 300 Project [8].

Injectivity threshold (1.5 kg/bar min), accordingly ATAP tests and bear out during Connectivity Test Inter-Wells, corresponds to the following parameter values:

- $WHP_i = 30–37$ bar

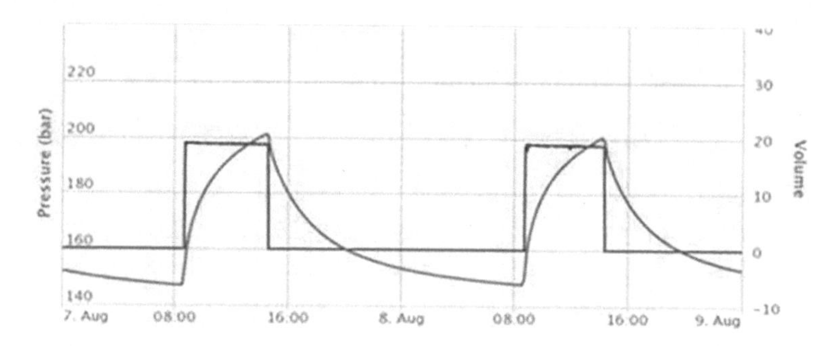

Fig. 15 Injection pulse test under WHP control mode (WHP = 60 bar, $t_{injection}$ = 7 h, $t_{fall\ off}$ = 17 h). Injected flow-rate (rectangular pulse) and BHP evolution (parabolic curve)

- Q_i = 60 kg/min.

According to this injectivity threshold, brine injection tests in pressure control mode were designed as:

- Pulse injection tests
- Continuous injection tests.

Injection tests under well head pressure control were carried out in the range WPH_i = 30 bar and WPH_l = 75 bar (WHT = 10–15 °C). Injection and fall off periods play a key role to analyze the hydraulic reservoir behavior for determining permeability and transmissivity. Time setting for these periods was as follows:

- Pulse injection tests, $t_{injection}$ = 7 h, $t_{fall\ off}$ = 17 h
- Continuous injection tests, $t_{injection}$ = 24–72 h, $t_{fall\ off}$ = 50–140 h.

Figure 15 shows the operating parameter evolution for a pulse test with set point WHP = 60 bar being needed a brine flow rate Q ~ 19 m^3/h to maintain constant well head pressure. During the injection period (7 h) BHP shows a parabolic evolution which does not reach an asymptotic scenario, what reveals a transient hydrodynamic period in the fracture network.

Injection pulse tests are easily performed and provide relevant information on data from the injection and fall off period for established periodicity. In any case, the injection periods are so short that they cannot be considered representative of an industrial injection, which is why longer tests must be carried out. The evolution of the injection is then analyzed with continuous tests, and in particular, to identify the seismic response threshold and the effects of injecting at pressures with values close to the LOT.

Figure 16 shows operating parameter evolution for a continuous injection test for a set-point WHP = 80 bar during 36 h. For this period BHP remained constant with a value of 220 bar, and decreasing evolution of flow rate (ranged between 38 and 20 m^3/h) as the fractures reached their maximum filling capacity (transient period).

As a complement to the tests in pressure control mode, others in flow control mode were designed by injecting brine for values between Q_i and Q_f, paying special

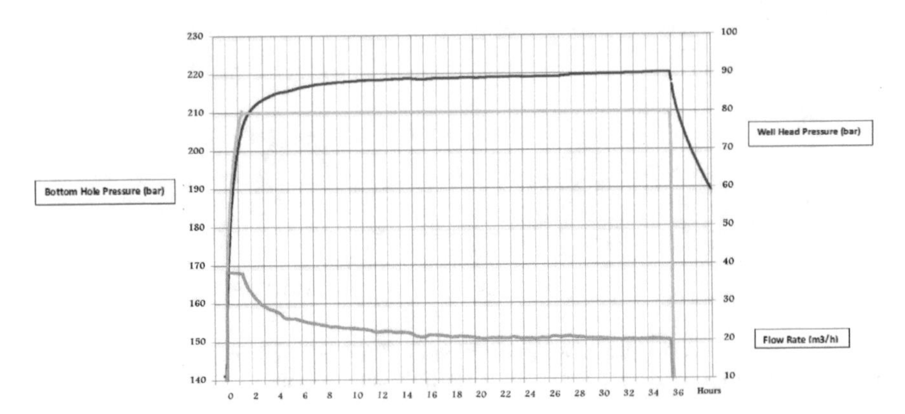

Fig. 16 Continuous injection test under WHP control mode (WHP = 80 bar, $t_{injection}$ = 36 h $t_{fall\ off}$ = 140 h). Injected flow-rate (green line), WHP and BHP evolution (olive and red lines)

attention to the values close to the fracture pressure of reservoir. Through these tests, it is intended to analyze the evolution of the injection pressure during a given period, and particularly, whether stabilization for mentioned parameter is achieved. The selected flow rate is usually the nominal design value of the CO_2 injection system at Hontomín pilot that is 120 kg/min.

Another type of test is the flow rate control mode with monitoring the evolution of well head pressure and with constant pressure at the bottom. In this type of tests, it is a question of checking the variation of the head pressure, keeping the pressure in the open hole constant by controlling the flow rate. The results obtained during the performance of several tests of this type for BHP values (from left to right in the graph) of 150, 155, 160, 175, 180 and 200 bar respectively, are shown in Fig. 17.

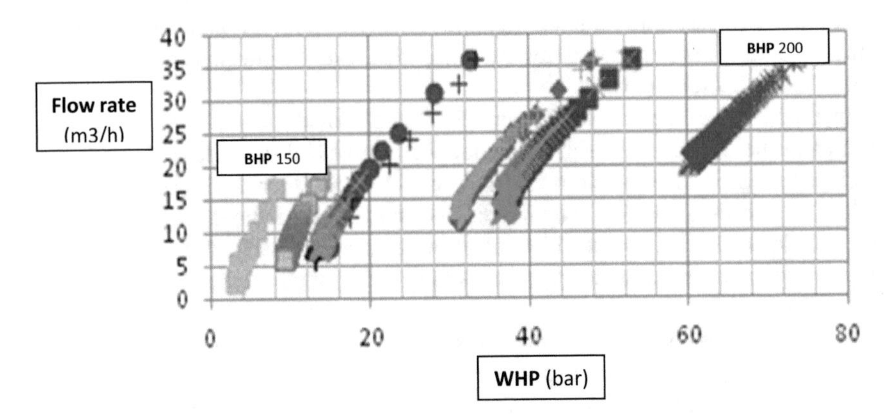

Fig. 17 Tests conducted in flow rate control mode with constant bottom pressure (BHP 250, 155, 160, 175, 180 and 200 bar)

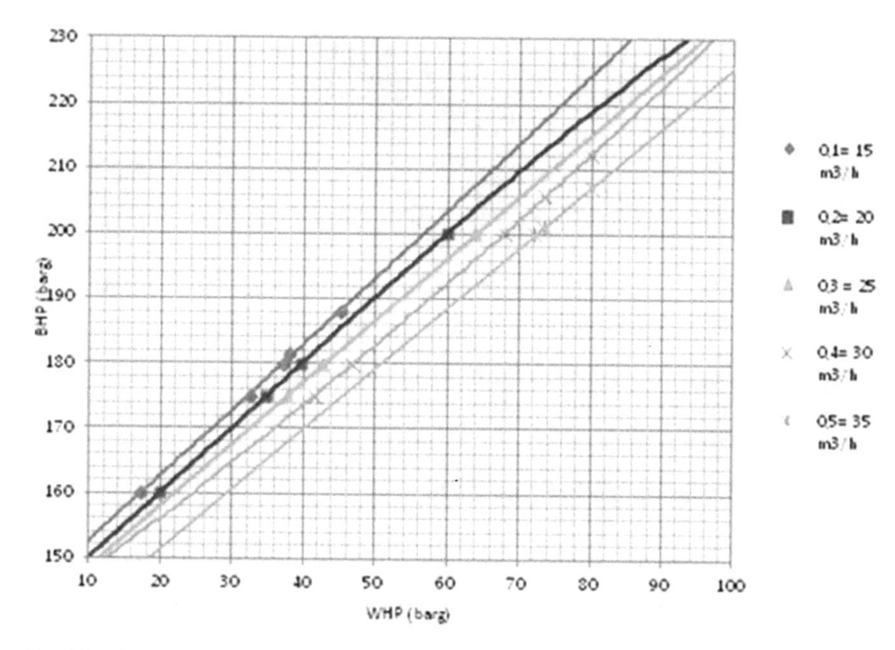

Fig. 18 Injectivity abacus

Data obtained from the development of the different tests described so far, and particularly from those corresponding to the flow rate control mode with constant bottom pressure, provide relevant information on operating conditions (WHP, BHP and flow rate values). The representation of their evolution may be graphed in form of injectivity abacus, as shown in Fig. 18.

Through the injectivity abacus, a preliminary design of the injection strategies can be drafted, having to correlate the values obtained with the brine injection tests and those corresponding to CO_2 injection. A relevant fact that must be taken into account in relation to the use of this abacus is the variation that it can experience during the project life. Therefore, the information provided corresponds to a certain period of the pilot operation, being necessary to conduct characterization tests from time to time that corroborate the absence of changes.

For the period from May to December 2014 a total amount of 14,000 m^3 of brine were injected in the following tests:

- 2 tests in flow rate control mode (Flow rate 120 kg/min)
- 2 tests in flow rate control mode (Flow rate 300 kg/min)
- 1 test in pressure control mode (WHP = 60 bar, t = 36 h)
- 28 short pulse tests in pressure control mode (WHP = 60 bar, t = 7 h)
- 1 test in pressure control mode (WHP = 80 bar, t = 36 h)
- 1 short pulse test in pressure control mode (WHP = 80 bar, t = 7 h)
- 2 short checking pulse tests in pressure control mode (WHP = 60 bar, t = 7 h).

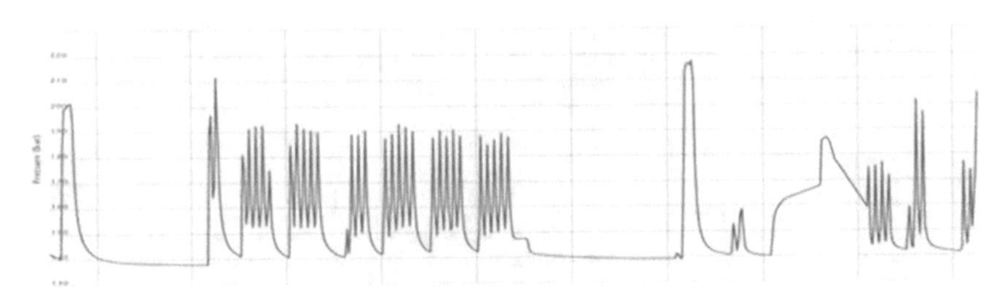

Fig. 19 Brine injection history showing BHP evolution for the period May–December 2014

Figure 19 shows the graph with the history injection of referred period.

6.2 *Co-injection of Brine and CO_2*

Regarding the results from ATAP tests and particularly the geochemical effects produced probably by ion migration [11] due to the co-injection of brine and CO_2 in supercritical phase, injection campaigns using blends of CO_2 (50%) and artificial brine (50%) were carried on site in order to enhance the secondary permeability in the carbonate fractures.

Tests were conducted in pressure control mode with a set point WHP = 75 bar and WHT = 10 °C to assure the CO_2 injection in liquid phase, thus a total amount of 1500 tons of carbon dioxide were injected. Afterwards, other 800 tons were injected to design safe and efficient strategies. A detailed description and analysis will be done about this last issue at Chapter "Safe and Efficient CO_2 Injection".

Figure 20 shows the evolution of pH of samples from resident brine taken with the U tube device (deep sampler) in pre and post injection situations, denoting an acidification thereof in the short term as it decreases from values of 7.3 to 6.5.

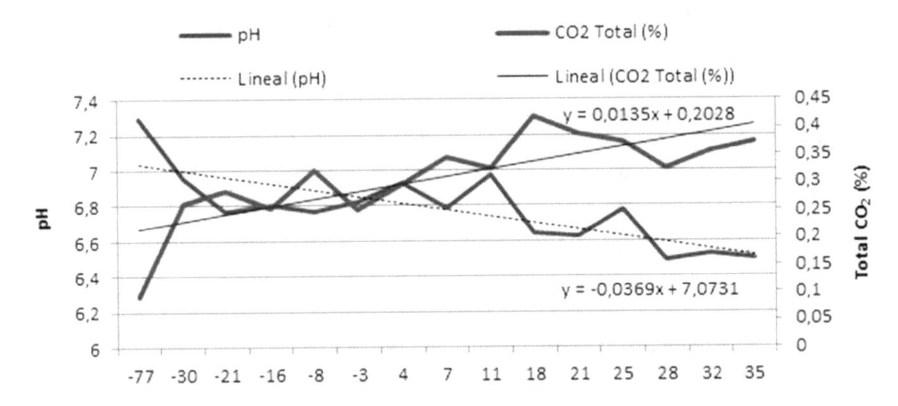

Fig. 20 CO_2 saturation and pH evolution in the resident brine of Hontomín

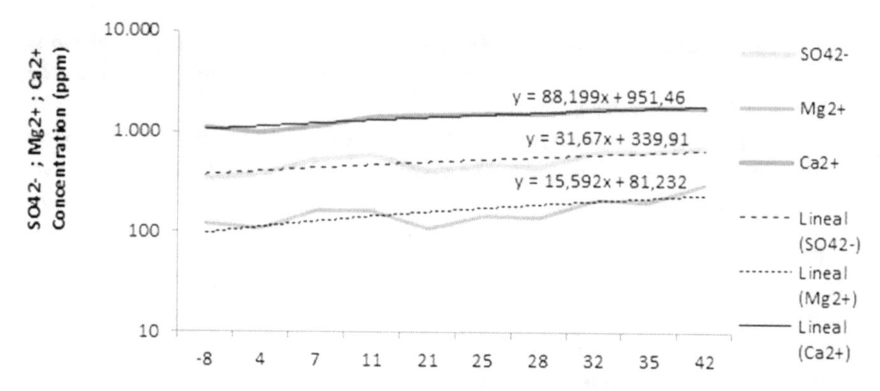

Fig. 21 Ion evolution in Hontomín resident brine during co-injection of CO_2 and artificial brine

Figure 21 shows the results of ion chromatography of referred samples, achieved under the conditions described above. A tendency to increase SO_4^{2-}, Mg^{2+} and Ca^{2+} ions can be observed as CO_2 has been injected in the reservoir.

These results suggest the reactivity between the two-phase flow of CO_2 and brine with the materials that fill the Sopeña Formation (reservoir) fractures, probably due to ionic migration, being consistent with the results provided by the ATAP test effluent analysis, described in Chapter "Laboratory Scale Works". The interpretation of test results will be carried out in Sect. 7.

Finally, a short pulse test in flow rate control mode (set point = 300 kg/min) was conducted, whose results were compared with the first one developed on site in order to assess the permeability increase in the fractures. The limit pressure at well bottom after 6 injecting hours was 206 bar in the initial test. The pressure reached at final test was considerably lower for the same period of time, reaching a value of 188 bar. The main conclusion is the effects of the injection at pressures close to LoT probably produced geomechanical changes in the fracture network and increased the permeability in the reservoir. On the other hand, the geochemical effects of co-injection of CO_2 and brine produced the alteration of fracture fillings, which also impacts on permeability increase [11].

Regarding the cap-rock integrity, pressure values recorded in the injection well (HI) and correlated with those recorded in the observation well (HA) did not reveal the existence of leakage pathways, neither hydrogeological network located at swallow aquifers detected any anomaly.

7 Modeling and Data Interpretation

The interpretation of data provided by the aforementioned characterization campaigns is not simple because of the complexity of the naturally fractured reservoir and their hydraulic behavior. Modeling used to analyze referred data was developed as follows:

1. Use of analytical code to interpret the results of the brine injections and determine preliminary values of reservoir hydraulic parameters, analyzing the effects that produces in the vicinity of the well
2. Results from the injection of blends of brine and CO_2 are interpreted by the history matching of the operating parameters with a numerical code that considers the hydrodynamic and geochemical effects of CO_2 injection in carbonate fractures.

Analytical codes are usually based on Darcy's Law, as a classic expression of fluid dynamics in a porous medium, and its linkage with the Navier–Stokes equation that determines the movement of a viscous fluid, considering flow rates of injection and the effective reservoir surface [27].

Figure 22 shows the evolution of pressure during brine injection and fall off phase (blue line) of permeability test 5, and simulated by Saphir™ [28] double porosity model (red line) that best matches with the behavior of well surrounding reservoir. The model provides the matching of the pressure and its derivative with respect to time.

Interpretation by modeling suggests that permeability would tend to improve in the vicinity of the well at a distance between 12 and 20 m from its axis, with a transmissivity of 24.2 mD m for a useful surface of 28 m and for a permeability of 0.86 mD and is estimated in the far zone from 3 to 5 mD.

Test described despite being the least representative because it has been carried out over the entire length of the open hole, yields more favorable values to describe the method used. The same procedure was followed to calculate the permeability of the rest of tests.

Saphir™ code was also used to interpret the connectivity tests inter-wells. Figure 23 shows the evolution of the pressure in pulses (brown line), described

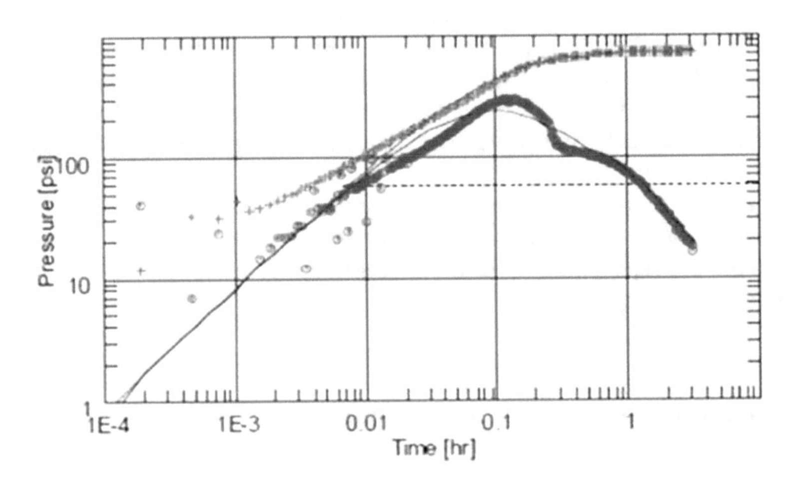

Fig. 22 PTFS n° 5 BHP evolution during injection and fall off phases (blue line) and modeled results (red line) using Saphir™

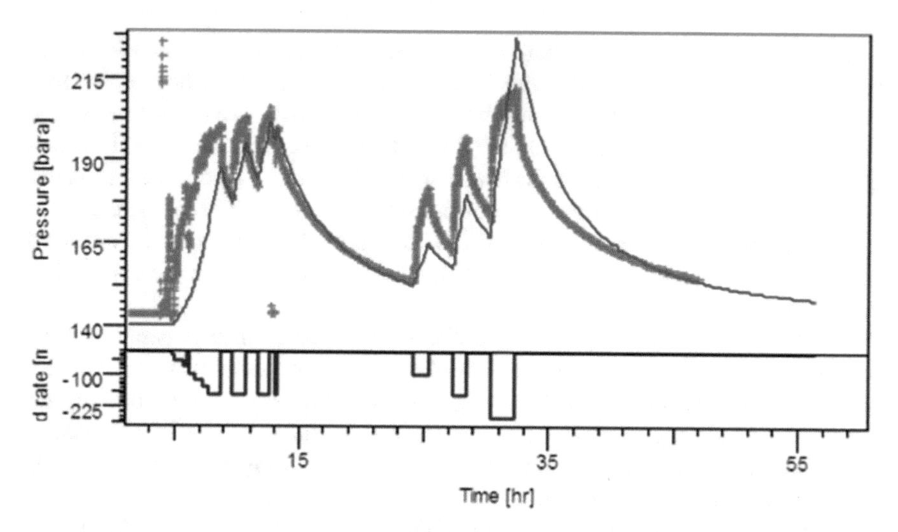

Fig. 23 CTIW BHP evolution during injection and fall off periods (brown line) and modeled results (red line) using Saphir™

in Sect. 5.2, and results provided by double porosity model (red line), estimating a permeability value of 0.86 mD in the vicinity of injection well (HI).

Regarding the history matching of data from all brine injection tests and modeled results provided by the Saphir™ double porosity model, the corresponding to the evolution of the pressure measured at the bottom of the injector well (HI) is in the upper part of Fig. 24 (green color) and also the model results (blue color). Pressure measured at the bottom of the observation well (green color) and the results of the model (red color) are shown in the lower part of the figure.

From the interpretation of results achieved with Saphir, an increase in the permeability in the fracture network is observed due to the effects of dilatency and extension of new fractures, from the 0.86 mD that were initially estimated during PTFS to the 5.4 mD calculated by modeling [11].

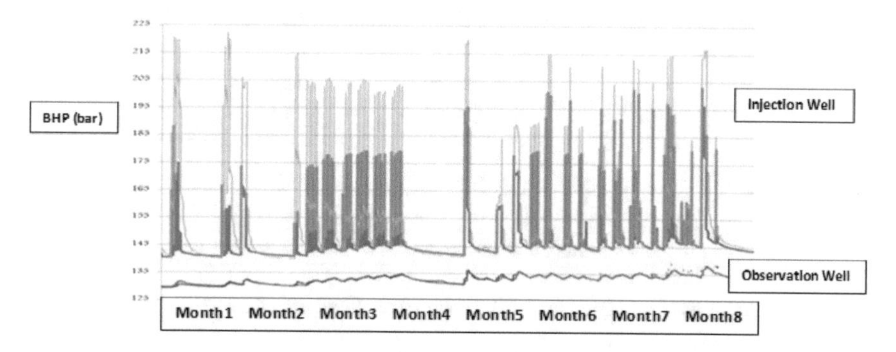

Fig. 24 Interpretation of brine injection tests by history matching using Saphir™

Saphir™ has relevant limitations since considers transport conditions in single-phase flow and for a medium with hydraulic transmissivity in the pore matrix, what means the results obtained must be verified through the use of more advanced codes [29]. On the other hand, Saphir™ does not consider the hydrodynamic effects produced by brine injection, which are usually associated with a non-linear response of double permeability aquifer. Moreover, effects produced by the chemical and thermal phenomena of co-injection of CO_2 and artificial brine must be analyzed and properly simulated. All these effects must be interpreted with the appropriate numerical code.

The code used to achieve these goals was GEM [30], which is based on state equations and simulates a multiphase flow of up to three phases and different components. This code is suitable for modeling of fractured reservoirs and especially for the geological storage of carbon dioxide, since it considers both effects related to hydrodynamic and geochemical phenomena that occur during the injection of CO_2 in Hontomín geological complex.

Figure 25 shows the history matching of the evolution of the bottom pressure and model results in the injection well (upper part) and in the observation well (lower part) for tests of co-injection of CO_2 and brine, analyzed in Sect. 6.2. Green curves correspond to the measured pressure values and the blue ones with the results of GEM model. Real values and model results match well in the injection periods, but it is necessary to perform a modeling update that improves the matching in fall off periods. Fracture permeability calculated is 15 mD, which represents an increase of almost 10 mD compared to the value contributed by Saphir during the brine injection phase which was 5.4 mD.

This permeability increase confirms the assumption that geochemical reactivity effects produced by the co-injection of brine and CO_2 have occurred in the reservoir as suggested by the pH analysis and ionic chromatography of resident brine [11].

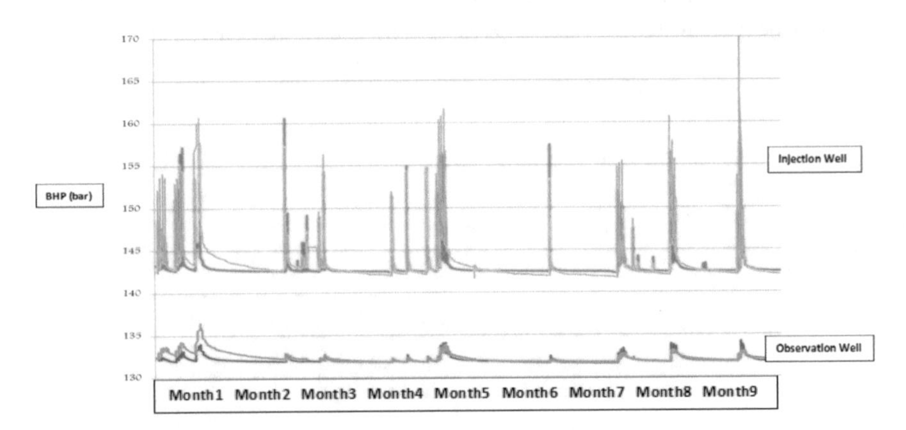

Fig. 25 Interpretation of co-injection tests by history matching using GEM™

8 Concluding Remarks

Main concluding remarks of on-site hydraulic characterization tests regarding the experience gained during campaigns conducted at Hontomín Technology Development Plant for CO_2 geological storage are the following:

- Both new wells drilled at Hontomín pilot for injection and observation (HI/HA) are located in the same geological block with fluid transmission between them
- Hydraulic properties of reservoir are typical of double porosity medium, with very poor primary permeability and fluid transmissivity dominated by fractures
- Brine injection tests produced a permeability increase in the fractures due to the effects of dilatency and extension of new fractures, from 0.86 mD that were initially estimated during PTFS to 5.4 mD calculated by modeling using Saphir™
- Co-injection of CO_2 and artificial brine induced new permeability increase due to geochemical reactivity effects, from 5.4 to 15 mD calculated by modeling using GEM™.

Main issue to discuss and further analyze is whether all phenomenon occurred and results achieved correspond to well bore effects in the vicinity of HI and HA, or they really represent the reservoir behavior. As mentioned above, following the procedure described in this chapter it is necessary to develop campaigns between the injection well and the rest of legacy wells, being of main challenges to overcome during the early injection planned in ENOS Project, in which, a total amount of 10,000 t of CO_2 are planned to inject on site [14].

Acknowledgements The experiences and results showed in this chapter form part of the project “OXYCFB 300” funded by the European Energy Program for Recovery (EEPR) and the Spanish Government through Foundation Ciudad de la Energía-CIUDEN F.S.P. Authors acknowledge the role of the funding entities, project partners and collaborators without which the project would not have been completed successfully.

This document reflects only the authors’ view and that European Commission or Spanish Government are not liable for any use that may be made of the information contained therein.

Glossary

ATAP Alta Temperatura y Alta Presión
BHP Bottom hole pressure
BHT Bottom hole temperature
BOP Blowout preventer
CT Computer tomography
CTIW Connectivity test inter wells
DAS Distributed Acoustic Sensing System
DTS Distributed Temperature Sensing System
ERT Electric Resistivity Tomography

HA Observation well
HDDP Heavy Duty Double Packer system
HI Injection well
LOP Leak off pressure
LOT Leak off test
MD Measured depth
OM Optical Microscopy
PTFS Permeability test at field scale
Q Flow rate
SEM Scanning Electrode Microscopy
$\mathbf{t_{fall\ off}}$ Fall off period
$\mathbf{t_{injection}}$ Injection period
WHP Well head pressure
WHT Well head temperature
XRD X-Ray Diffraction
XRF X-Ray Fluorescence

References

1. Chadwick, A., Arts, R., Bernstone, C., May, F., Thibeau, S., & Zweigel, P. (Eds.). (2008). *Best practice for the storage of CO_2in saline aquifers-observations and guidelines from the SACS and CO2STORE projects* (Vol. 14). British Geological Survey. Available online https://core.ac.uk/download/pdf/63085.pdf. Accessed November 5, 2019.
2. Gupta, N. (2008). *The Ohio River valley CO_2storage project. AEP Mountaineer Plan* (Final technical report). West Virginia: Battelle Columbus Operations. Available online https://digital.library.unt.edu/ark:/67531/metadc929044/. Accessed November 5, 2019.
3. Mishra, S., Kelley, M., Zeller, E., Slee, N., Gupta, N., Battacharya, I., & Hammond, M. (2013). Maximizing the value of pressure monitoring data from CO_2 sequestration projects. *Energy Procedia, 37*, 4155–4165. Available online https://doi.org/10.1016/j.egypro.2013.06.317. Accessed November 5, 2019.
4. Finley, R. J., Frailey, S. M., Leetaru, H. E., Senel, O., Couëslan, M. L., & Scott, M. (2013). Early operational experience at a one-million tonne CCS demonstration project, Decatur, Illinois, USA, *Energy Procedia, 37*, 6149–6155. Available online https://doi.org/10.1016/j.egypro.2013.06.544. Accessed November 5, 2019
5. Worth, K., White, D., Chalaturnik, R., Sorensen, J., Hawkes, C., Rostron, B., et al. (2014). Aquistore project measurement, monitoring, and verification: From concept to CO_2 injection. *Energy Procedia, 63*, 3202–3208. Available online https://doi.org/10.1016/j.egypro.2014.11.345. Accessed November 5, 2019.
6. Wilson, M., & Monea, M. (2004). *IEA GHG Weyburn CO_2monitoring & storage project* (Summary report 2000–2004). Available online https://ieaghg.org/docs/general_publications/weyburn.pdf. Accessed November 5, 2019.
7. Liu, H., Tellez, B. G., Atallah, T., & Barghouty, M. (2012). The role of CO_2 capture and storage in Saudi Arabia's energy future. *International Journal of Greenhouse Gas Control, 11*, 163–171. Available online https://doi.org/10.1016/j.ijggc.2012.08.008. Accessed November 5, 2019.

8. *The Compostilla Project «OXYCFB300» carbon capture and storage demonstration project. Knowledge sharing FEED report*. Global CCS Institute publications. Available online https://hub.globalccsinstitute.com/sites/default/files/publications/137158/Compostilla-project-OXYCFB300-carbon-capture-storage-demonstration-project-knowledge-sharing-FEED-report.pdf. Accessed November 5, 2019.
9. Rubio, F. M., Garcia, J., Ayala, C., Rey, C., García Lobón, J. L., Ortiz, G., & de Dios, J. C. (2014). Gravimetric characterization of the geological structure of Hontomín. In *8ªAsamblea Hispano-Lusa de Geodesia y Geofísica*, Évora.
10. Spane, F. A., Thorne, P. D., Gupta, N., Jagucki, P., Ramakrishnan, T. S., & Mueller, N. (2006). Results obtained from reconnaissance-level and detailed reservoir characterization methods utilized for determining hydraulic property distribution characteristics at Mountaineer AEP 1. In *PROCEEDINGS, CO2SC Symposium*, Lawrence Berkeley National Laboratory, Berkeley, CA, March 20–22, 2006. Available online https://www.osti.gov/servlets/purl/881621#page=163. Accessed November 13, 2019.
11. de Dios, J. C., Delgado, M. A., Marín, J. A., Salvador, I., Álvarez, I., Martinez, C., & Ramos, A. (2017). Hydraulic characterization of fractured carbonates for CO_2 geological storage: experiences and lessons learned in Hontomín Technology Development Plant. *International Journal of Greenhouse Gas Control, 58C*, 185–200.
12. Alcalde, J., Marzán, I., Saura, E., Martí, D., Ayarza, P., Juhlin, C., et al. (2014). 3D geological characterization of the Hontomín CO_2 storage site, Spain: Multidisciplinary approach from seismic, well-log and regional data. *Tectonophysics, 627*, 6–25.
13. Le Gallo, Y., & de Dios, J. C. (2018). Geological model of a storage complex for a CO_2 storage operation in a naturally-fractured carbonate formation. *Geosciences, 8*(9), 354–367. Available online https://doi.org/10.3390/geosciences8090354. Accessed November 13, 2019.
14. *Enabling on-shore CO_2storage in Europe*. EC Horizon 2020. Available online https://www.enos-project.eu/
15. de Dios, J. C., Delgado, M. A., Marín, J. A., Martinez, C., Ramos, A., Salvador, I., & Valle, L. (2016). Short-term effects of impurities in the CO_2 stream injected into fractured carbonates. *International Journal of Greenhouse Gas Control, 54*, 727–736.
16. Gastine, M., Berenblyum, R., Czernichowski-Lauriol, I., de Dios, J. C., Audigane, P., Hladik, V., et al. (2017). Enabling onshore CO_2 storage in Europe: Fostering international cooperation around pilot and test sites. *Energy Procedia, 114*, 5905–5915.
17. Le Gallo, Y., de Dios, J. C., Salvador, I., & Acosta Carballo, T. (2017). Dynamic characterization of fractured carbonates at the Hontomín CO_2 storage site. In *EGU General Assembly Conference 2017*. EGU2017-3468-1 (Vol. 19, p. 3468).
18. Accelerating CCS Technologies. Available online https://www.act-ccs.eu/
19. de Dios, J. C., & Martínez, R. (2019). The permitting procedure for CO_2 geological storage for research purposes in a deep saline aquifer in Spain. *International Journal of Greenhouse Gas Control, 91*, 102822. Available online https://doi.org/10.1016/j.ijggc.2019.102822
20. de Dios, J. C., Álvarez, I., & Delgado, M. A. (2018). *Laboratory procedures for petrophysical characterization and control of CO_2geological storage in deep saline aquifers*. Spanish CO_2 Technology Platform (PTECO2) Publications. Available online https://www.pteco2.es/es/publicaciones/procedimientos-de-laboratorio-para-la-caracterizacion-petrofisica-y-el-control-de-almacenes-geologicos-de-co2-en-acuiferos-salinos-profundos
21. Valle, L. (2012). ATAP design of a device for dynamic and static petrophysical studies of interaction between rock-brine-supercritical CO_2 in deep saline aquifers. In *GERG Academic Network Event*, Brussels (Belgium), June 14–15. Available online https://www.gerg.eu/publications/academic-network-2012. Accessed November 26, 2019.
22. Valle, L. (2014). Hontomín reservoir condition tests. In *IV Spanish-French Symposium on CO_2Geological Storage*, May 13–14, 2014.
23. SOLEXPERTS AG. *Company profile*. Available online https://www.solexperts.com. Accessed November 28, 2019.

24. Li, G., Lornwongngam, A., & Roegiers, J. C. (2009). *Critical review of leak-off test as a practice for determination of in-situ stresses.* ARMA-09-003. American Rock Mechanics Association. Available online https://www.onepetro.org/conference-paper/ARMA-09-003. Accessed December 5, 2019.
25. Vilarasa, V., Olivella, S., Carrera, J., & Rutqvist, J. (2014). Long-term impacts of cold CO_2 injection on the cap rock integrity. *International Journal of Green House Gas Control, 24*, 1–13. Available online https://doi.org/10.1016/j.ijggc.2014.02.016. Accessed December 5, 2019.
26. Wiese, B., Böhner, J., Enachescu, C., Würdemann, H., & Zimmermann, G. (2010). Hydraulic characterisation of the Stuttgart formation at the pilot test site for CO_2 storage, Ketzin, Germany. *International Journal of Greenhouse Gas Control, 4*, 960–971.
27. Dake, L. P. (1998). Fundamentals of reservoir engineering.
28. *Saphir pressure transient analysis.* KAPPA Engineering. Available online https://www.kappaeng.com/software/saphir/overview. Accessed January 20, 2020.
29. de Dios, J. C., Le Gallo, Y., & Marín, J. A. (2019). Innovative CO_2 injection strategies in carbonates and advanced modeling for numerical investigation. *Fluids, 4*(1), 52. Available online https://doi.org/10.3390/fluids4010052
30. Berkeley Lab. (2016). Available online https://ipo.lbl.gov/lbnl1613/. Accessed January 20, 2020.

Safe and Efficient CO_2 Injection

Alberto Ramos, Carlos Martínez, and J. Carlos de Dios

Abstract CO_2 injection must be safe assuring the integrity of seal-reservoir pair for long-term gas trapping, and it must be also efficient as commercial activity that seeks business profit. Injection in tight reservoirs, as fractured carbonates, usually needs high pressure values to reach proper injectivity ranges, what means the geological complex integrity could be put at risk due to rock fracturing or fault-slip that may generate leakage pathways. The study case of OXYCFB300 Project is addressed in this chapter, analyzing how the CO_2 injection in dense state is conducted in the naturally fractured reservoir of Hontomín pilot. Inputs from hydraulic characterization tests developed on site, addressed on Chapter "On-Site Hydraulic Characterization Tests", are briefly analyzed, tackling how the injection strategies were designed under mentioned safety and efficiency criteria. Spanish Patent "*Industrial process for CO_2injection in dense state from pipeline transport condition to permanent geological trapping*" is analyzed in the chapter. Particularly, the innovative process with alternative injection of CO_2 and brine, as well as, the necessary mitigation tools to preserve the operation safety. Both measures to control the bottom pressure build up and the effect of impurities existing on CO_2 stream are analyzed. Finally, the conditions necessary to become a patent in operation are described, as future works planned in the project.

Keywords Safe and efficient injection · CO_2 dense phase · Alternative injection of CO_2 and brine · Bottom pressure build-up · Effects of impurities

---

A. Ramos (✉) · C. Martínez
School of Mines and Energy, Technical University of Madrid, Calle de Rios Rosas 21, 28003 Madrid, Spain
e-mail: alberto.ramos@upm.es

J. C. de Dios
Foundation Ciudad de la Energía-CIUDEN F.S.P., Avenida del Presidente Rodríguez Zapatero, 24492 Cubillos del Sil, Spain

J. C. de Dios et al. (eds.), *CO_2 Injection in the Network of Carbonate Fractures*, Petroleum Engineering, https://doi.org/10.1007/978-3-030-62986-1_5

1 Introduction

CO_2 injection is the final stage of Carbon Capture and Storage (CCS) chain, what involves the gas pumping from surface to a deep geological complex for permanent trapping. Thus, CO_2 is injected in different reservoirs such as depleted hydrocarbon sites, deep saline aquifers and coal beds, inter alia geological formations. In all cases, reservoir management is necessary what implies pressure and capacity control [1].

Injectivity is probably one of most relevant operating parameters, since it correlates the flow rate injected on site and the pressure necessary. The ideal setting for the operator is a high permeable reservoir that implies low pressure for gas injection at the largest possible rates into the smallest number of wells. Nevertheless, CO_2 injection at constant pressure on permeable sites do not assure a constant flow rating because of the multiphase flow effects and the disturbance produced by the injection in the resident brine surrounding the well vicinity [2].

In fractured carbonates, such as Hontomín case, injectivity is usually low due to the poor primary permeability and fluid transmission dominated by fractures [3], what involves a high anisotropy degree of fluid movement that hinders CO_2 plume tracking. Other relevant effect are the geomechanical changes induced by the high pressure necessary to inject, which implies the strength-strain state alteration of seal-reservoir pair or even the fault-slip [4] generated on site.

According to the stated above, CO_2 injection must be safe, what involves the bottom pressure build up in the seal-reservoir pair must reach a value to preserve the geological complex integrity, avoiding leakage pathways generated by rock fracturing or fault-slip which put at risk the long-term gas trapping [5]. In the same way, the injection must be efficient as industrial operation, and for this purpose dense state is most appropriate to inject CO_2 at the largest possible rates in each well. This issue depends on gas transport conditions and mainly the reservoir depth that determines the bottom pressure, and therefore, the well head pressure to inject CO_2.

This chapter address the early CO_2 injection strategy carried out at Hontomín Technology Development Plant (TDP) in the framework of OXYCFB300 Project [6], funded by the European Energy Program for Recovery (EEPR) and the Spanish Government. Main goal of 1st phase of the Project was the technology development for CO_2 oxy-combustion capture, inland transport and geological storage in saline aquifers, supporting a future demo 300 Mw OXY CCS fired coal power plant, located in the existing area of Compostilla Power Plant at Cubillos del Sil (León) in the northwest of Spain. 2nd phase of the Project was focused on the construction of the new 300 Mw CCS Power Plant, the 135 km on-shore transport pipeline and Duero storage site located close to Sahagún village (León). Unfortunately, final investment decision was negative due to the existing constrictions of Spanish electricity market in 2013 and the 2nd phase did not go forward. Project results from 1st phase were the Power Plant Front End Engineering Design (FEED), Duero site exploration, confirming its ability to become an on-shore commercial site for CO_2 geological storage, and the following pilot plants operated by Foundation Ciudad de la Energía-CIUDEN F.S.P, which cover the full CCS chain:

- 30 Mw OXY Capture Circulating Fluidized Bed Boiler
- 3 km transport closed-loop test rig
- On-shore Hontomín Technology Development Plant for storage.

Research activities carried out at Hontomín pilot would provide "real life" experiences on characterization, operation and monitoring to Duero site and other on-shore storage sites, as foreseen in ENOS Project [7]. Therefore, OXYCFB300 Project conditions for the injection are determined by the transport operating parameters and the reservoir depth in Duero site. CO_2 transport by in-land pipeline was planned in supercritical phase and the maximum reservoir depth was close to 2500 m, therefore, CO_2 injection in dense state seems to be the most appropriate. Hontomín TDP is equipped with surface tanks to store CO_2 in cryogenic conditions, injection pumps, CO_2 thermal conditioning and one injection well (HI) and other monitoring well (HA) that reach 1600 m depth [8], what means dense CO_2 injection can be done.

To assure dense state, the alternative injection of brine and CO_2 is carried out, known as water alternating gas (WAG). Thus, previously gas injection the well tubing is filled with pressurized brine, and after CO_2 injection this operation is repeated as well sealing. Injection process is under the Spanish Patent ES 201500151 "*Industrial process for CO_2injection in dense state from pipeline transport condition to permanent geological trapping*" which is analyzed in this chapter.

The impurities play a key role since condition the injection efficiency as they impact on operating parameters, and produce corrosion in transport and storage equipment. The effects and mitigation tools are described and analyzed, as good practice guidelines for operators. Likewise, as future works planned in the project the conditions necessary to become patent in operation are shown.

2 Safety and Efficiency Criteria During Injection

Main goal of CO_2 geological storage is the permanent trapping of the injected gas in the seal-reservoir pair. The concept "permanent" means that CO_2 must be confined for hundreds of years in the structural complex avoiding its release to the atmosphere. This issue is not trivial and requires control measures along the whole project (i.e. operation, monitoring and site closure). Therefore, the safety criteria used to design the injection strategies are focused on the operation to avoid gas leakages during this phase and subsequently because of seal-reservoir pair is put at risk or well completion is damaged.

Even in most adequate storage formations, CO_2 could leak due to the buoyancy of the separate phase gas, the induced pressure gradients from injection and the variable nature of strata serving as barriers to upward migration [9], and therefore monitoring and mitigation measures are necessary. It is not in our power to avoid some effects intrinsically related to the nature and tectonics of geological formations or to thermodynamic and chemical properties of the gas. However, other effects such as the induced by pressure gradients from injection can be identified and mitigated using the proper operating procedure.

Bottom pressure build-up control is essential activity within reservoir management. Pressure increase can lead to cap-rock failure in extreme cases, what would mean the generation of critical leak pathways what does not ensure the trapping integrity. In some study cases, such as tight reservoirs with dual transmission in the porous matrix and fractures, injection may induce new fractures which cannot reach the seal [10] to extend the plume migration. These effects are usually accompanied by micro-seismicity generation, which may be a problem of public acceptance mainly in on-shore sites. Some operating measures can be taken to prevent these impacts, such as *EPA Geologic CO_2Sequestration Technology and Cost Analysis* that establishes to limit the injection pressure to 90% of fracture pressure of injection formation [11].

Pressure increase induced by CO_2 injection may produce changes on the strain–stress state of seal-reservoir pair that could generate tensile fracturing and shear slip along pre-existing fractures. This scenario needs a coupled reservoir-geomechanical analysis to estimate the maximum pressure value without causing the breach [12]. Modeling simulations must predict the bottom pressure build up along the injection period according to the injectivity and maximum amount of CO_2 envisaged. Figure 1 shows the modeled scenario with reservoir pressure values at Hontomín pilot after 50 kilotons of CO_2 are injected on site with a yearly rate of 10 kilotons/year.

We have to take in mind that most probably leakage pathways in addition to the faults are the wells, particularly those ones dedicated to injection. Regarding the high pressure necessary to inject CO_2 in dense state, well equipment is subjected to high efforts that could put at risk the safety of the operation. Likewise, CO_2 and brine is a corrosive blend whose impact is enhanced due to the presence of some impurities (e.g. SO_2, NO_X, H_2S, inter alia acid gases). Therefore, special materials to prevent or minimize corrosion caused by carbon acid must be used during well equipment, taking special care for well cementing by the use of specialty cements in remediation or new well construction. Likewise, Operation & Maintenance Plan for CO_2 injection must include a surveillance protocol that envisages monitoring

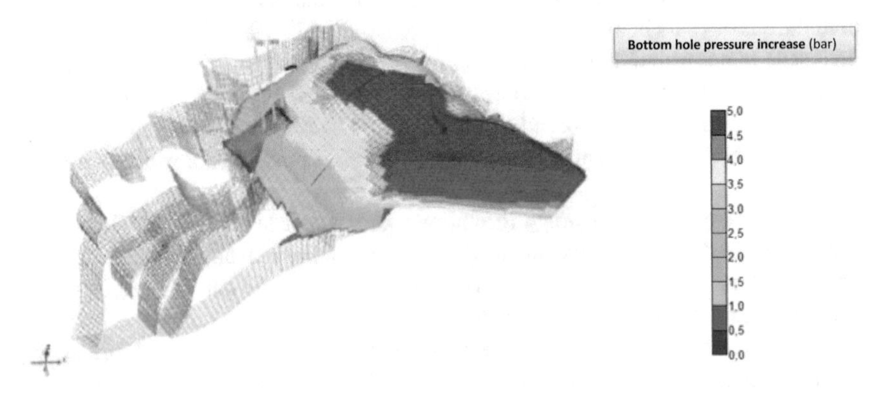

Fig. 1 Final bottom pressure at Hontomín site after 50 kilotons of CO_2 injected on site

and prevention through mechanical integrity tests to control pressure effects on well completion, the corrosion of internal and external parts of well heads, manifolds, piping and monitoring devices and periodic surveys of cement integrity [11].

Efficiency criteria are focused to reach the largest possible CO_2 injection rates into the smallest number of wells, as mentioned above. To achieve this goal, some challenges must be overcome such as the surface equipment needed for the injection (e.g. pumps, compressors, temporary storage tanks, thermal conditioning equipment, inter alia components) or the type and dimensions of well completion, mainly the injection well head and tubing. However, dense phase during injection is the key for operation efficiency.

Figure 2 shows CO_2 phases according to the pressure and temperature. There are four phases: gas, liquid, supercritical and solid. Triple point (−56.6 °C and 5.18 bar) corresponds to the state where gas, liquid and solid phases co-exist. Following the saturation line between liquid and gas phases, the critical point is reached for pressure of 73.8 bar and 31.1 °C of temperature. For higher temperatures and pressures CO_2 is in supercritical phase, which is the hybrid of liquid and gas.

Supercritical CO_2 is the phase for deep geological storage because of the high values of confining pressure and temperature. Depending on these parameters, CO_2 density in the reservoirs is in the range 0.15–0.95 t/m^3 for supercritical phase. Obviously, the greater reservoir depth the higher stored CO_2 amount due to the increase of pressure and temperature that induce higher density in storage.

Taking into account this constrain that limits the maximum density of CO_2 stored on site up to 0.95 t/m^3, dense gas state seems the most appropriate according to the

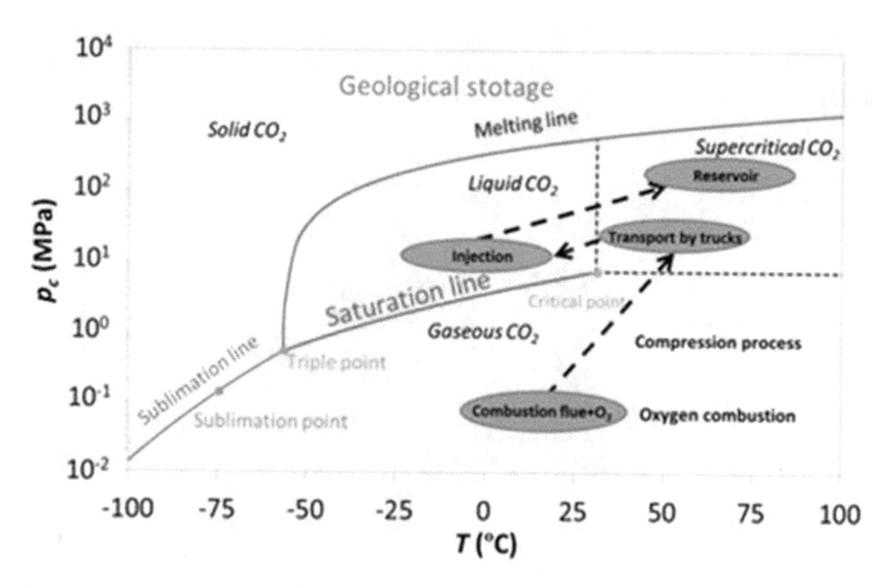

Fig. 2 Carbon dioxide pressure–temperature diagram

efficiency criteria for carbon dioxide injection. Dense state involves those pressure and temperature values corresponding to supercritical CO_2 at higher densities and the liquid phase.

3 Case Study: Project OXYCFB300

As mentioned above, Hontomín TDP was one of three pilots for covering CCS technology chain, designed, constructed and commissioned under OXYCFB300 Project framework [6]. Hontomín experiences were intended to lead and support the injection operations at Duero site, which was explored by ENDESA electric company in the project. Several activities at Hontomín were planned in support to the industrial scale activities to be developed by ENDESA, being a main goal the investigation of technologies required by the CO_2 injection in the reservoir and its long-term monitoring.

The OXYCFB300 "Compostilla" Project was a Carbon Capture and Storage (CCS) integral commercial demonstration project, including oxy-combustion capture, inland transport and geological storage in saline aquifers, supporting a future demo 300 Mw OXY CCS fired coal power plant, located in the area of existing Compostilla power plant operated by ENDESA in the province of León, northwest of Spain [6]. Unfortunately, FID was negative due to the existing constrictions of Spanish electricity market in 2013.

CO_2 transport was planned by a pipeline 136 km length which would join the capture plant with the industrial storage site located at Sahagún (León), known as Duero site. Figure 3 shows the locations of capture and storage sites where they were planned to be constructed, and Hontomín Pilot.

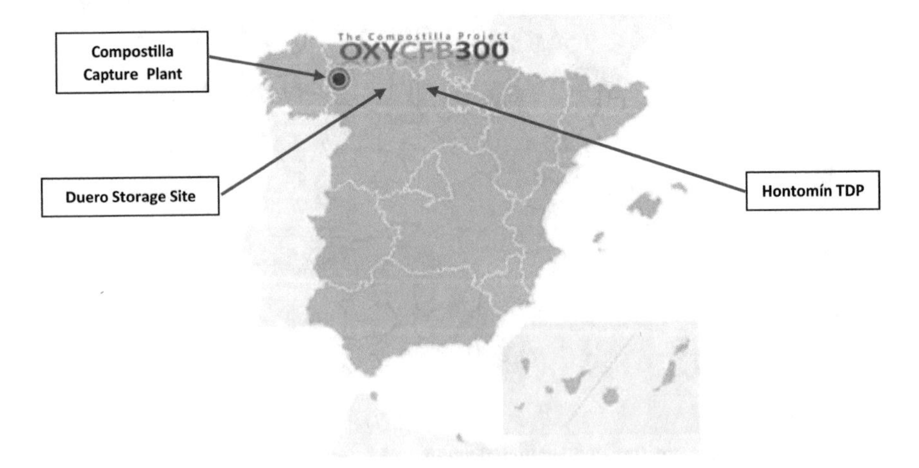

Fig. 3 Location of capture plant, industrial storage site and Hontomín TDP in OXYCFB300 Project

Table 1 OXYCFB300 pipeline transport

OXYCFB300 Project transport pipeline	
Pipeline inlet pressure (bar)	150
Pipeline diameter (in.)	14
Pipeline length (km)	136
Pipeline inlet temperature (°C)	10

Table 2 OXYCFB300 injection conditions

OXYCFB300 Project injection conditions	
Injection wells	2 + 1
Monitoring wells	2 + 3
Well head pressure (bar)	60–80
Injection rate (kg/s)	23–35
Well head temperature (°C)	10

In order to provide realistic data for industrial operation at Duero site, injection strategies carried out at Hontomín were designed using the transport operating parameters for in-land pipeline and the reservoir conditions. Table 1 shows main characteristics and operating parameters of the pipeline transport planned in OXYCFB300 [6].

Regarding Duero site conditions, injection would reach 2500 m depth in the reservoir, according to the figures showed in Table 2 [6].

Regarding these data, CO_2 injection was planned in dense state.

4 Input from Hontomín Hydraulic Characterization

As mentioned in Chapter "On-Site Hydraulic Characterization Tests", the design of safe and efficient injection strategies is the main goal of site hydraulic characterization, taking into account CO_2 characteristics and the property distribution of seal reservoir pair where gas will be permanently trapped.

Structural complexity of Hontomín was the main challenge to carry out the hydraulic characterization of the site. It is a dome with the reservoir (Sopeña Formation) and seal (Marly Liassic and Pozazal Formations) located at the depth from 900 (top of the dome) to 1832 m (flanks). The reservoir is composed of naturally fractured carbonates, limestones in its upper part and dolostones at the bottom (see Fig. 1 of Chapter "Light Drilling, Well Completion and Deep Monitoring"), where primary permeability is very poor and fluid transmission is mainly dominated by fractures. As matrix effective porosity is low, the injectivity in this dual medium was conditioned by the fracture network being lower than expected. This fact faced at laboratory scale and field tests was the main challenge to overcome during hydraulic characterization [3].

Table 3 Permeability evolution in porous matrix and fractures during Hontomín hydraulic characterization

Permeability	Field permeability tests	Brine injection	Co-injection CO_2 + brine
Fractures (mD)	0.866	5	15
Matrix (mD)	Negligible	Negligible	2

The injection of 14,000 m^3 of artificial brine and the co-injection of CO_2 + brine subsequently, induced geomechanical changes in the fracture network and chemical reactivity with fracture fillings which produced the permeability increase from 0.866 mD measured during first hydraulic field tests to 15 mD at the end of characterization process [3]. Table 3 shows the permeability evolution during hydraulic characterization process.

The reader has more detailed information about mentioned tests, results and their interpretation in Sects. 5, 6 and 7 of Chapter "On-Site Hydraulic Characterization Tests".

Brine injection carried out in pressures close to Leak off Test (LoT) (i.e. equal or higher than 75 bar) induced geomechanical changes in the fracture network, such as dilatency and new fractures, which enhanced secondary permeability. This fact was underpin by CO_2 + brine co-injection that produced a second step of permeability increase because of chemical reactivity mainly between the acidified brine and fracture fillings. Co-injection also produced a slight increase of matrix permeability through the porous media, but far away from usual values to assure the CO_2 trapping.

Taking into account both boundary conditions established in OXYCFB300 Project, addressed in precedent section, and the final results from Hontomín Hydraulic characterization reveal CO_2 injection must be conducted in high pressure (i.e. equal or higher than 75 bar) to assure the dense gas state along the tubing and the fracture opening for plume migration within the tight reservoir. This necessity conditions the site safety, as mentioned in Sect. 2, since the operation could produce micro-seismicity around the on-shore pilot, being able to generate public opposition to this technology, and the cap-rock breakup which could be even worse as the complex integrity is put at risk. These issues are addressed on Sect. 7 where mitigation tools for the operational risks are described and analyzed.

5 Spanish Patent ES 201500151

CO_2 dense state at the well head (i.e. pressure, flow-rate and temperature) is conditioned by the depth and geomechanical, geothermal and geochemical conditions of seal-reservoir pair where carbon dioxide will be permanently trapped. Therefore, mass flow-rate from transport pipeline to the storage formation is assured by the adjustment of transport conditions to the corresponding of well head injection.

Stability of fluid mass flow requires that gas state changes, with relevant density alterations, must be avoided, which would ensure the integrity and long-life of injection wells. For this purpose, well head pressure must be hold in a constant value of 80 bar during CO_2 injection, and subsequently, the existing gas in the well must be pushed to the reservoir, ensuring its trapping and avoiding a potential leakage.

Spanish Patent "*Industrial process for CO_2injection in dense state from pipeline transport condition to permanent geological trapping*" tackles how to inject CO_2 in an efficient a safe manner, accomplishing the requirements mentioned above.

5.1 *Injection Process*

Patented industrial process involves the following steps:

1. Well pressurization using brine
2. CO_2 conditioning to be injected
3. CO_2 injection
4. Well head and tubing cleaning.

Well pressurization using artificial or resident brine is mandatory to hold the pressure along the well, and particularly its head, close to the value that assures CO_2 injection in dense state. This maneuver must be conducted prior to injection or because of some interruption occurs, such us breakdowns, scheduled maintenance shutdowns or high induced pressure that lead to well shut-in and wait and see how to continue the operation.

If the well is not adequately pressurized prior to CO_2 injection with values usually lower, which could reach the atmospheric pressure in some cases, harmful effects may happen since the pressure of the gas coming from the pipeline would be much greater than well head pressure. Transient effects are produced by the adiabatic gas expansion in this cases, such as the well cooling which usually decreases injectivity due to hydrates formation [13], and even the tubing plugging can take place in extreme cases. Moreover, CO_2 expansion in the tubing or near the well bore region, with lower viscosity than the corresponding to transport conditions, noticeably increases gas velocity possible causing erosion or cavitation in the lines.

Figure 4 shows the well head equipment of the injection well (HI) at Hontomín pilot, with CO_2 and brine lines and the corresponding valves.

Well pressurization starts opening both Xmas tree valves (i.e. CO_2 and brine valves), and subsequently the injection pump operates according to the set-up values of pressure or flow-rate established in the control system. Then brine is injected through the backflow and Xmas tree valves reaching the well head and the injection tubing, while the backflow valve installed at the end of CO_2 line avoids the brine entrance on gas injection piping. Well head pressure increases proportionally to the injected flow rate and reservoir permeability, from the initial value existing previously the injection to the final value of 80 bar. Flow rate is adjusted to reach the final pressure avoiding to significantly exceeding the set value which would mean

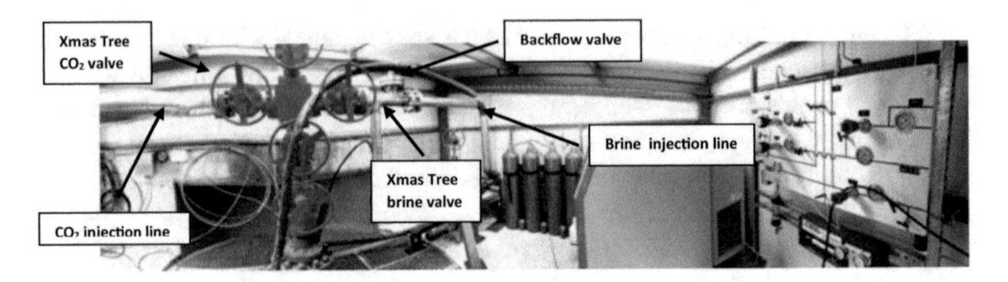

Fig. 4 Well head equipment of injection well at Hontomín TDP

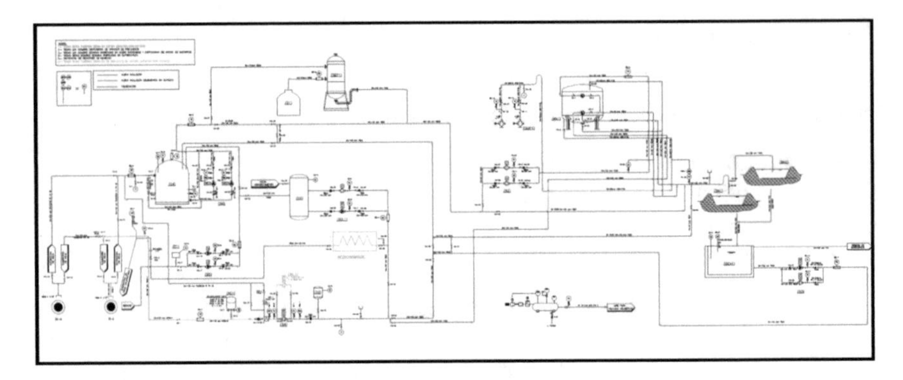

Fig. 5 PI&D of brine injection facility at Hontomín TDP

increasing excessively the pressure to inject CO_2. Figure 5 shows the brine injection facility at Hontomín TDP.

Once the injection well is pressurized with brine, CO_2 injection starts while brine flow rate is decreasing sequentially until reaches zero. The injection is controlled to achieve the pressure in the range between 80 bar and the corresponding transport value, set in 150 bar for the study case (see Sect. 3), and temperature equal or higher than 10 °C. The backflow valve installed at the end of the brine line (see Fig. 4) avoids the entrance of carbon dioxide in the piping causing corrosion.

As field tests were conducted injecting CO_2 stored in cryogenic tanks (i.e. 20 bar and −20 °C) at Hontomín (see Fig. 6), thermal conditioning was necessary in addition to the pressure increase by using injection pumps. If the mentioned operational requirements apply to transport by pipeline, as planned in OXYCFB300 Project, this case would entail the need for a surface equipment to adjust CO_2 pressure to a value closer to 80 bar, and also the temperature (i.e. $\geq$10 °C) which plays a key role to avoid hydrates formation [14] and the harmful effects they produce in well lines, as mentioned above.

Hereinafter CO_2 injection will be carried out according to the control mode selected to operate, preserving the bottom pressure build up to safe values, as described and analyzed in Sect. 7. Following control modes were implemented at Hontomín Technology Development Plant:

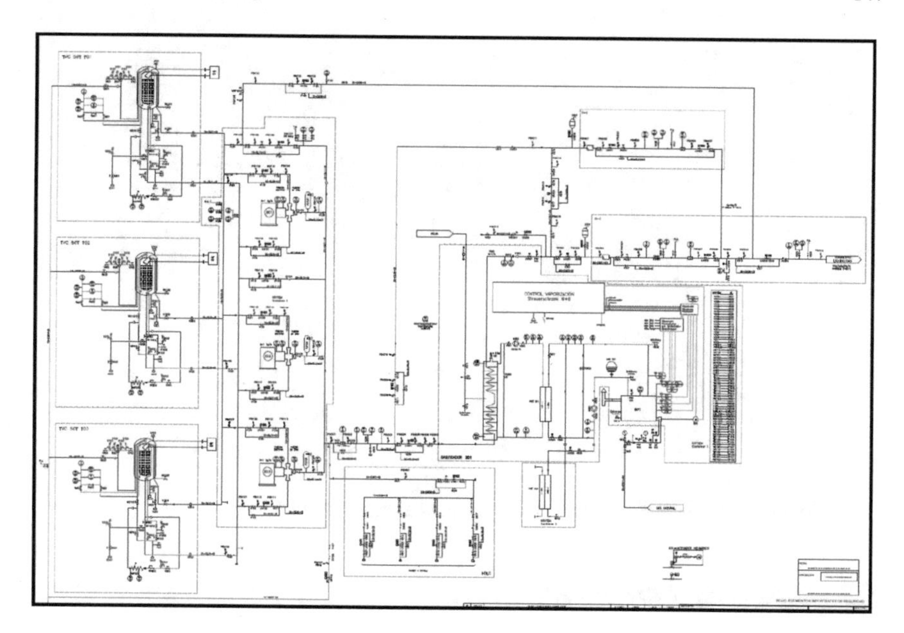

Fig. 6 PI&D of CO_2 injection facility at Hontomín TDP

- Pressure control
- Flow rate control.

The injection facility operates in pressure mode using this parameter as control variable, establishing a set-point slightly higher than 80 bar. Thus, the injection equipment will adjust CO_2 flow rate to reach the planned well head pressure. On the other hand, flow rate mode uses this parameter as control variable during the operation. Special care has to be taken using this last mode since well head and bottom hole pressures must reach appropriate values regarding safety criteria for the injection. More detailed description on injection control modes is given in Sect. 5.2.

Finally, when CO_2 injection is necessary to stop for any reason that suggests the gas remove from the well head and rest of lines according to operational safety criteria, brine injection starts again as described above, while CO_2 flow rate decreases to reach zero. From this moment, brine pumping will be conducted up to reach an injected volume double to corresponding of well head and rest of injecting lines. Once this brine amount is injected, both Xmas Tree valves corresponding to CO_2 and brine injection lines will be closed, checking whether well head pressure decreases to reach at least 50% of set-point. If this pressure state is achieved, Xmas Tree valve opens and brine injection starts again to push the remaining CO_2 existing in the hydraulic column to the reservoir.

Figure 7 shows the operating phases of patented injection process described above.

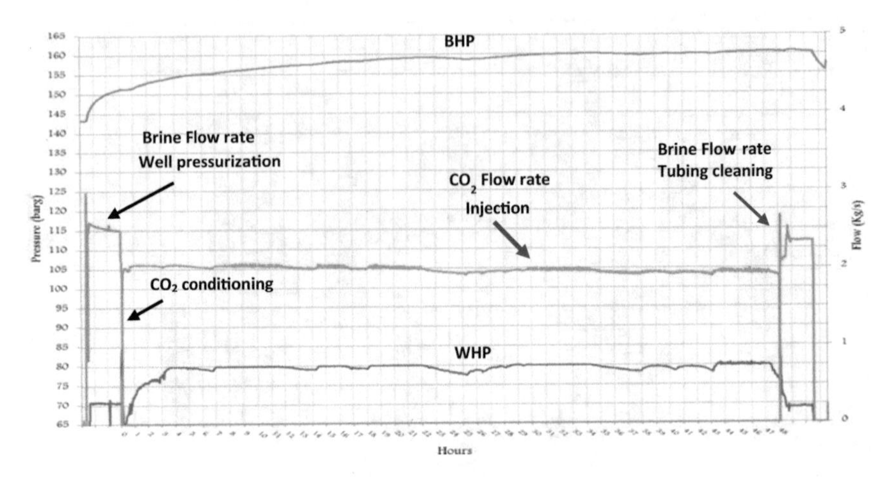

Fig. 7 Steps of CO_2 injection process under Spanish Patent ES 201500151

5.2 *Operating Parameters*

Main operating parameters to control CO_2 injection process as described in precedent section are the following:

- Well head pressure (WHP)
- Bottom hole pressure (BHP)
- Well head temperature (WHT)
- Bottom hole temperature (BHT)
- CO_2 flow rate
- Brine flow rate.

The injection conducted under pressure control uses WHP as operational monitoring parameter, whose set-point is 80 bar as mentioned above. Figures 8 and 9 show the evolution of operating parameters during 24-h CO_2 injection test in pressure control mode.

WHP (light blue line in Fig. 8) maintains constant during the test period except the beginning and end that correspond to well pressurization and tubing cleaning with brine respectively. WHT (orange line in Fig. 8) is also surrounding the set-point of 10 °C during the injection. For the fall off period both parameters evolve according to the atmospheric pressure and environmental temperature. BHP (marine blue and red lines in Fig. 8, corresponding to both P/T sensors located at bottom hole of the injection well (HI), see Fig. 2 of Chapter "Light Drilling, Well Completion and Deep Monitoring") increases from 140 to finally 160 bar reached when test finished, which does not accord with WHP value hold during the injection. This issue is analyzed and discussed at Sect. 7, where mitigation tools for operational risk prevention are described.

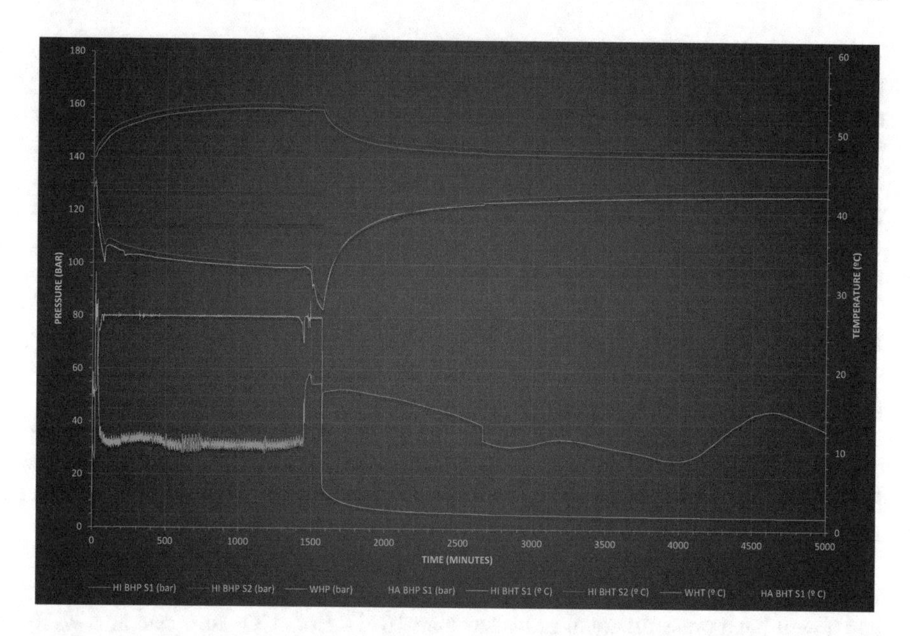

Fig. 8 Evolution of WHP, BHP, WHT and BHT during 24-h CO_2 injection test in pressure control mode

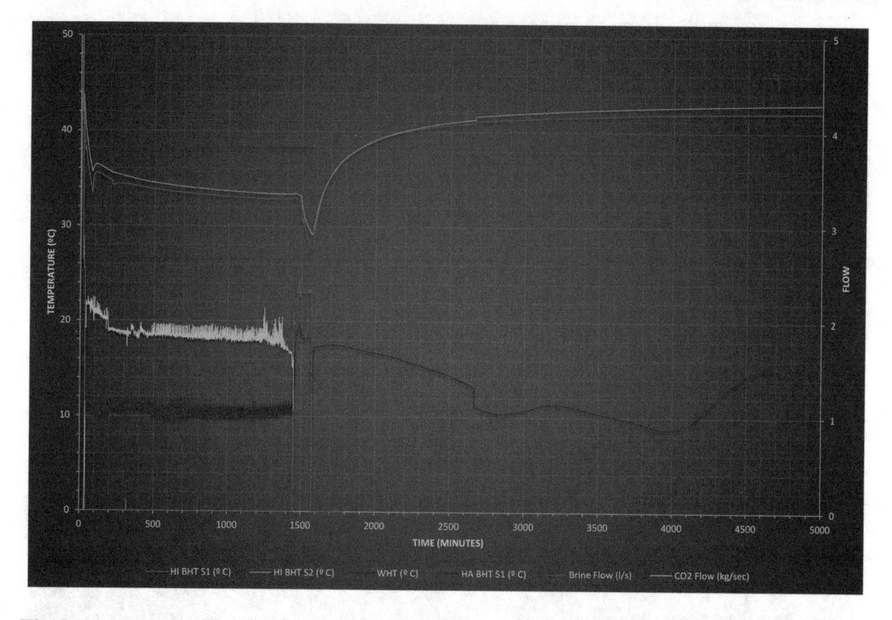

Fig. 9 Evolution of BHT, WHT, CO_2 and brine flow rates during 24-h CO_2 injection test in pressure control mode

Bottom hole pressure and temperature measured by the sensor located in the cap rock (see Fig. 2 of Chapter "Light Drilling, Well Completion and Deep Monitoring") at the observation well (HA) (dark blue and brown lines in Fig. 8) are constant along the test what proves the integrity of seal-reservoir pair.

Regarding BHT evolution in the injection well (orange and purple lines in Fig. 9), the trend is homogeneous except at the beginning and end of the test where a relevant decrease appears, corresponding to brine injection (olive line in Fig. 9). This effect is because of brine flow rate is much greater than corresponding to CO_2 injection, and depending on environmental temperature at the surface where store water pools are located, BHP decrease is more or less intense. Nonetheless, this relevant temperature change at the bottom hole must be studied in the future, particularly, how this effect could impact on geomechanical changes at the reservoir and cap rock.

CO_2 flow rate evolution (light green line in Fig. 9) shows a slightly decrease during 24-h injection, while WHP maintained constant (i.e. 80 bar). This means the injectivity also decreases during the test, addressing in Sect. 6 the possible reasons and implications for a future industrial process.

Figure 10 shows the evolution of operating parameters during 8-h CO_2 injection in flow rate control mode. A set-point of 2 kg/s was established for the operation and a well head pressure equal or higher than 10 °C. Both CO_2 flow rate and WHP (light green and brown lines respectively in Fig. 10) maintain constant during the test, except the beginning and end as explained above, and an intermediate case

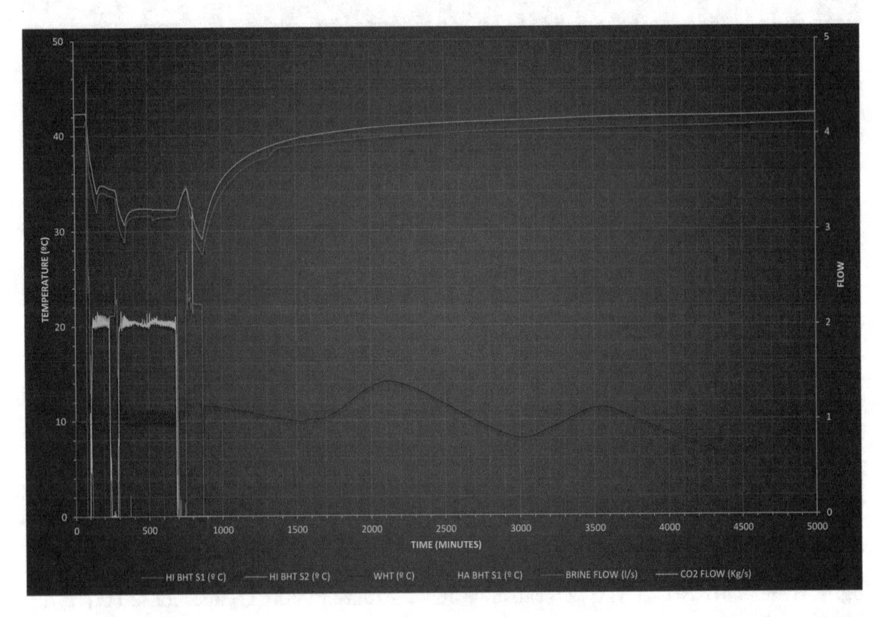

Fig. 10 Evolution of BHT, WHT, CO_2 and brine flow rates during 8-h CO_2 injection test in flow rate control mode

corresponding to a breakdown in the pumping system. This event suggested the cleaning of the tubing with brine and CO_2 injection re-start as shown in the graph.

The relevant temperature change continuous at the bottom hole induced by the brine injections.

6 Which Brine Alternating Gas Injection Entails

Brine alternating gas injection is a crucial operation in the process described in precedent sections, firstly, because it is the way for assuring the pressurization to reach the set-point to inject CO_2 in dense state, and the well completion clean up whether an interruption occurs and CO_2 remove is necessary for safety reasons. On the other hand, the brine column in the well ensures an adequate sealing in case the gas injection shut-in is prolonged, being needed an appropriate monitoring process for the decision making as part of the contingency plan.

Nonetheless, brine can produce harmful effects on the wells and surface equipment mainly due to the corrosion induced by the blend with the carbon dioxide and because of its high salinity degree [15]. Therefore, special care must be taken to select suitable materials to equip the wells, mainly those corresponding to injection lines, well heads, tubing, packers and monitoring devices, as well as, the casings and type of cement used for well completion, as mentioned above. Moreover, a surveillance protocol must be included as part of the Operation & Maintenance Plan of the injection facility, to check the mechanical integrity of the well components and surface equipment [11].

Brine alternating gas injection may also change operating parameters, such as the injectivity that depends on pressure and flow rate. This effect has been already studied in EOR Projects, where water alternating gas injection (WAG) altered the reservoir injectivity in porous media [16]. Taking into account the data from 24-h CO_2 injection test in pressure control mode, described in Sect. 5.2, values of main operating parameters are shown in Table 4.

Therefore, injectivity decreases close to 23% from initial value. Figure 11 shows the injectivity trend along the test, what reveals the flow rate decrease to maintain constant WHP. A possible cause of this effect could be the expansion of CO_2 and brine in the fracture network that produces the multiphase flow hysteresis [17]. Moreover, fracture behavior on CO_2 plume migration plays a key role, particularly, the geomechanical properties and their changes induced by the injection [3], which can also affect the injectivity change. Thus, Discussion is focused on whether these results

Table 4 Operating parameters of 24-h CO_2 injection test in pressure control mode

Parameter	Initial value	Final value
WHP (bar)	80	80
WHT (°C)	10	10
Flow rate (kg/s)	2.2	1.7

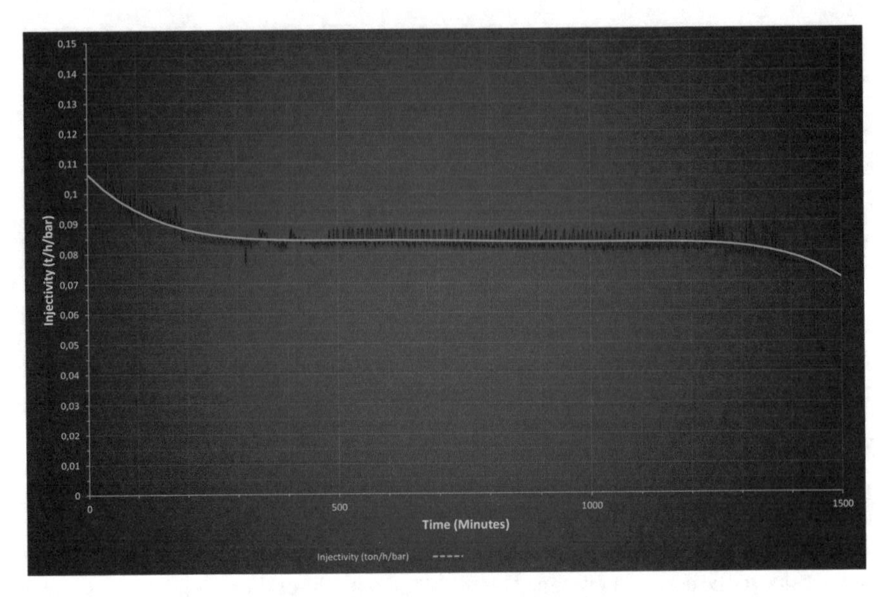

Fig. 11 Injectivity evolution during 24-h CO_2 injection test in pressure control mode

correspond to wellbore effects in the short-term, or they set a trend on long-term behavior of pair seal-reservoir. New injections conducted in similar conditions during several days are necessary to give proper solutions to the injectivity changes, BHP recovery term and BHP/T evolution, in order to analyze how alternating injection impacts on these operating parameters.

7 Mitigation Tools for the Operational Risks

Regarding the conditions established in OXYCFB300 Project and the safety criteria to develop CO_2 injection process, main operational risks are those related to BHP build up and the impacts of impurities existing in the carbon dioxide stream on the geological complex properties and the mechanical wholeness of well equipment.

BHP build up is critical to preserve the integrity of seal-reservoir pair, avoiding the generation of leakage pathways and micro-seismicity effects which could put the project viability at risk, as mentioned in precedent sections. On the other hand, the impurities existing in the captured flue gas usually induce corrosion on surface equipment and well completion, alter the geological properties because of geochemical reactivity or produce inefficiencies in the industrial operation.

Therefore, mitigation tools must be envisaged to carry out CO_2 injection in an efficient and safely manner.

7.1 How to Preserve Safe Pressure Values

The challenge at Hontomín was to assure the effective CO_2 injection preserving the integrity of seal-reservoir pair and avoiding micro-seismicity effects which could impact on public acceptance of CCS technologies. For that purpose, a non-commercial drop valve was designed by the engineering team of CIUDEN and Technical University of Madrid, to be anchored into the injection tubing at 1000 m depth to control the BHP build up.

Pressure drop in a long-length choke is determined by Darcy-Weisbach equation:

$$\Delta p = f \cdot \frac{L}{d_0} \cdot \frac{v_f^2}{2}$$

- Δp: Pressure drop (Pa)
- f: Friction coefficient
- L: Choke length (mm)
- d_0: Choke inner diameter (mm)
- v_f: Fluid velocity (m/s).

Friction coefficient depends on fluid circulation regime into the choke. Circulation regime may be laminar or turbulent which is determined by Reynolds number:

$$Re = \frac{v_f \cdot d_0}{\upsilon}$$

- Re: Reynolds number
- υ: Fluid kinematic viscosity (m^2/s).

Darcy-Weisbach equation as a function of mass flow rate, neglecting the losses in the transition joints in the pipe, is as follows:

$$\Delta p = K \cdot q_f^2$$

- K: Head losses factor (kg m^{-7})
- qf: Fluid mass flow rate (kg/s).

Graphs of Fig. 12 illustrates the simulation of CO_2 injection with the drop pressure valve $\Delta p = 60$ bar and flow rate of 2 kg/s installed on the tubing landing nipple located at 1000 m depth inside the injection well (HA) at Hontomín site. The operational parameters simulated vs depth are the following:

- Pressure along the well length: tubing pressure (red line) and hydrostatic pressure (green line)
- Fluid velocity (red line)
- Fluid density (red line)
- Tubing temperature (red line).

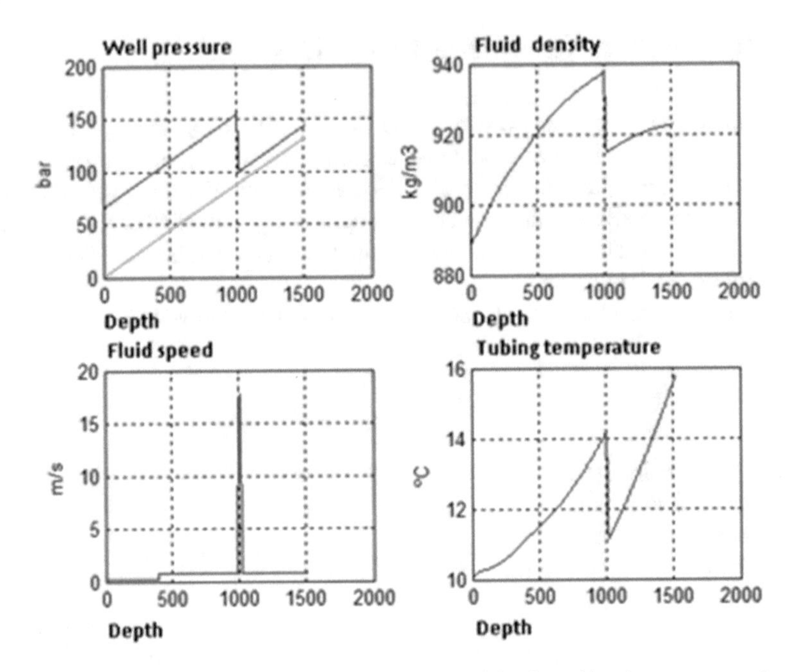

Fig. 12 Simulation of operational parameters during CO_2 injection with a drop pressure valve ($\Delta p = 60$ bar) installed in the landing nipple located 1000 m depth inside the tubing of HI well. Well pressure, fluid speed, density and temperature evolution depending on the depth

Graphs on the left side of Fig. 12 show the pressure decrease at the choke exit and velocity increase along its length. Regarding $\Delta p = 60$ bar for a flow rate of 2 kg/s and WHP of 75 bar, maximum overpressure at the bottom hole would be 15 bar which assures a safe injection that avoids the generation of new fractures, since reservoir LoT is in the range 60–70 bar (see Sect. 5.3 of Chapter "On-Site Hydraulic Characterization Tests"), and micro-seismicity events. Velocity increase along the choke ensures a fluid circulation regime according to the operating procedure existing at Hontomín TDP. Graphs on the right side show the density decrease at the choke exit produced by the pressure drop that is close to 2%, which does not significantly affect the operating efficiency, and the temperature decrease produced by Joule-Thompson effect [18]. This thermal phenomenon is relevant to ensure CO_2 injection in dense state, as will be discussed later.

Field tests were necessary to check simulation results and validate the use of the choke which has the following technical characteristics:

- Maximum Pressure: 20 MPa
- Flow rate: 0–2.25 kg/s
- Fluid temperature: 283–303 K
- Piping
    1. Inner diameter: 12.54 mm

Fig. 13 Choke installation into the HI injection tubing

2. Schedule: 40 s
3. Material: Super-duplex steel
4. Total length: 16–20 m.

Figure 13 shows the slick-line maneuver to install the long-length choke in the injection well (HI) at Hontomín TDP.

Table 5 shows average values of choke tests injecting CO_2 during Hontomín hydraulic characterization. Total length necessary for a pressure drop of 60 bar was 20 m, injecting a flow rate of 2.2 kg/s with a WHP of 80 bar. These figures are quite similar to those ones simulated. Nevertheless, well head pressure necessary to inject the maximum flow-rate is 6.6% higher than expected.

Figure 14 shows the thermal profile along the injection well (HI), where CO_2 temperature is the blue line and the corresponding to the existing water in the annular space between the tubing and casings is the green line. Data were measured by Distributed Temperature Sensor System (DTS) [19] which is an optic fiber anchored to the injection tubing (see Fig. 2 of Chapter "Light Drilling, Well Completion and Deep Monitoring").

Table 6 shows DTS technical characteristics, being the spatial resolution the distance over which the system responds to a step change in temperature. On the other hand, the temperature resolution is defined as the standard deviation of the

Table 5 Average figures of choke test

WHP (bar)	Flow-rate (kg/s)	ΔP choke (bar)	BHP overpressure (bar)
80	2.2	60	20

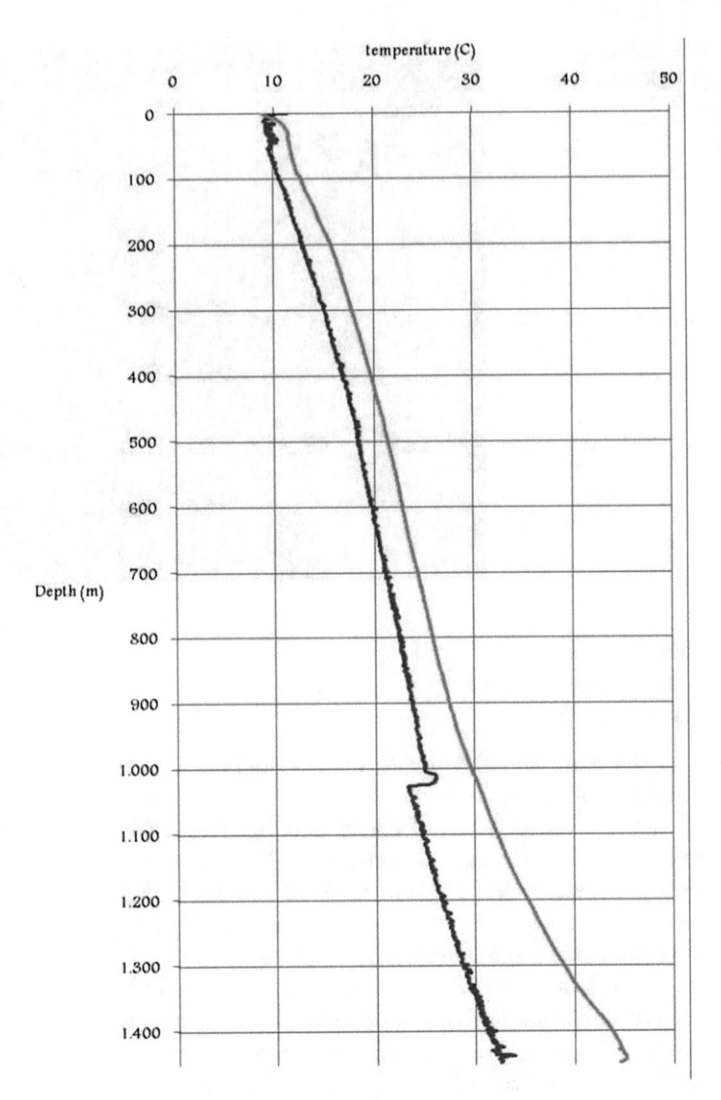

Fig. 14 Thermal profile along the injection well using DTS system. CO_2 injection (blue line) water in annular space (green line)

Table 6 DTS technical characteristics

Feature	Performance
Measurement range	0–5 km
Temperature range	0–150 °C
Temperature accuracy	0.1 °C
Minimal spatial resolution	40 cm
Minimal sampling resolution	12.5 cm

temperature measured by series of consecutive data points in distance, with the fiber maintained at a constant temperature over that distance.

Temperature decrease produced by Joule-Thompson effect at the choke exit is similar than the value predicted by simulation, although for higher temperature since simulation did not consider the value at the well head (i.e. 10 °C). Regarding this data and the measured pressure decrease, the installed choke as drop pressure valve in the injection tubing works properly to prevent a high BHP build-up which could put at risk the operation safety. Moreover, thermal profile proves also CO_2 injection was conducted in dense state.

7.2 The Effects of Impurities in the CO_2 Stream

Captured CO_2 stream wraps up other gases considered as impurities that impact on the safety and/or efficiency of injection. Their existence depends mainly on the composition of the fuel used, the chemical reactions associated to the combustion process where the flue gas is generated and the CO_2 capture method (e.g. pre-combustion, post-combustion and oxy-combustion) [20].

Main impurities existing in CO_2 stream captured by oxy-combustion technology in OXYCFB300 Project are: O_2, N_2, H_2O, CO, H_2S, SO_2 and NO_x. Tests at lab and field scale were carried out by Foundation Ciudad de la Energía (CIUDEN) in the framework of IMPACTS Project [21] to study the effects of impurities on carbon dioxide transport and storage.

Plug flow by lab dynamic tests were carried out to analyze the effects of SO_2 within CO_2 stream on Hontomín reservoir samples, described and analyzed by de Dios et al. [22]. Main results induced by short-term effects were the rock porosity change induced by the mixture of sulfur dioxide in the resident brine and the chemical reactivity by ion migration with a relevant pH decrease in the effluent.

Field tests were conducted by the co-injection of CO_2 + artificial air ($O_2 + N_2$) to study the effects of these impurities on the operation efficiency and reservoir capacity for CO_2 trapping. For a synthetic air concentration of 5.1%, based on BHP and BHT at the injection well (i.e. 158 bar and 31 °C), the WHP necessary to inject CO_2 stream increased 12.5% from the value used to inject pure carbon dioxide (i.e. from 80 to 90 bar) for a WHT of 10 °C and flow rate of 2 kg/s, as shown in Fig. 15. BHP was not affected during the testing period. It was also proved that CO_2 was stored in supercritical conditions, but with a relevant density decrease close to 8% (i.e. from 0.840 to 0.775 t/m^3) [22].

Taking into account the main results from IMPACTS Project [21], the short-term effects of impurities condition both efficiency and safety of industrial injection of CO_2, altering the properties of seal-reservoir pair that determine the complex permeability and its integrity, and also the operational injectivity and trapping capacity.

The mitigation tools applicable to this scenario are closely related to the CO_2 capture technology used at the emitter source, including specific process to remove

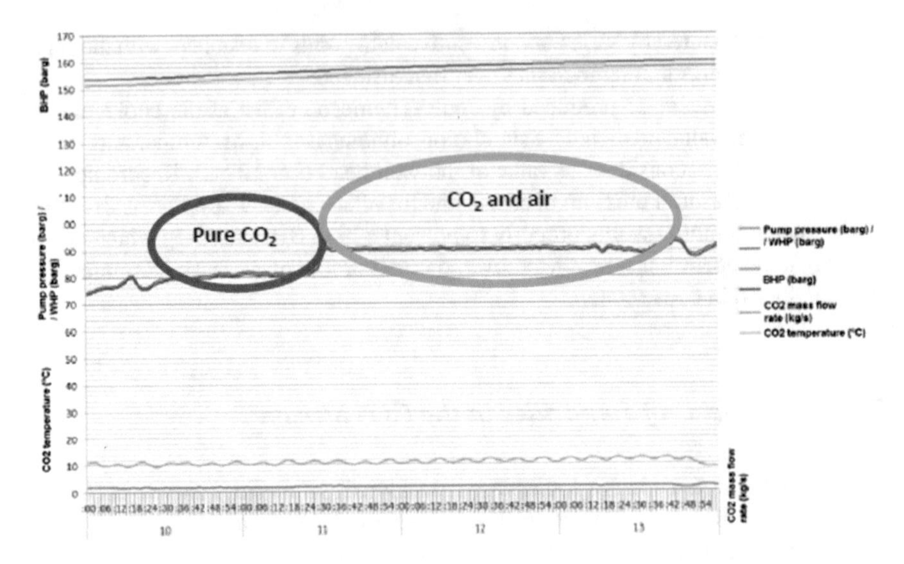

Fig. 15 WHP/BHP evolution for the co-injection of CO_2 + artificial air (5.1% O_2 + N_2)

certain impurities from the flue gas, and a surveillance protocol included in the O&M Plan, that envisages the monitoring and prevention through mechanical integrity tests to control the corrosion of internal and external parts of surface equipment and well completion, as mentioned in Sect. 2.

8 Concluding Remarks

Main concluding remarks on the industrial process to inject CO_2 in dense state from pipeline transport condition to permanent geological trapping are the following:

- CO_2 injection must be safe, avoiding the leakage pathways generated by rock fracturing or fault-slip which put at risk the long-term gas trapping. In the same way, the injection must be efficient, being carbon dioxide dense state the most appropriate to inject gas at the largest possible flow rates.
- Injectivity is the most relevant operating parameters, since it correlates the flow rate injected on site and the pressure necessary. Thus, the ideal setting is a reservoir that implies low pressure for gas injection at the largest possible amount into the smallest number of wells.
- OXYCFB300 Project is the study case that established CO_2 pipeline transport conditions (i.e. pressure/temperature of 150 bar and $\geq$10 °C, respectively), to inject the carbon dioxide in a deep saline aquifer 2500 m depth, with a WHP in the range 60–80 bar, which assures the gas dense state injection.
- Spanish patented process entails the following steps: 1. Well pressurization using brine to reach the injection set-point (i.e. 80 bar), 2. CO_2 conditioning to be

injected (i.e. 80 bar/$\geq$10 °C), 3. CO_2 injection, 4. Well head and tubing clean-up using brine.

- Two operating control modes were used for Hontomín injection tests, pressure and flow rate control modes. Each one uses the well head pressure and flow rate as control parameter, in the ranges 60–80 bar and 1–2.2 kg/s respectively.
- Brine alternating gas injection is a crucial operation because it is the way for assuring the pressurization to reach the set-point to inject CO_2 in dense state, the well completion clean up and brine column ensures an adequate well sealing in case the gas injection shut-in is prolonged.
- Brine alternating gas injection may change the operating injectivity. A possible cause of this effect could be the expansion of CO_2 and brine in the fracture network that produces the multiphase flow hysteresis. Nevertheless, geomechanical properties of fractures, and the changes induced by the injection, can also affect the injectivity change.
- Future CO_2 injections must be carried out at Hontomín TDP to give proper solutions to the injectivity change, BHP recovery term and BHP/T evolution, in order to analyze how alternating injection impacts on these operating parameters.
- Bottom hole pressure build up must be properly controlled for safe injection. For that purpose, a non-commercial drop valve (choke) was designed to be anchored into the tubing at 1000 m depth. Total choke length necessary for a pressure drop of 60 bar was 20 m, injecting a flow rate of 2.2 kg/s with a WHP of 80 bar.
- Main results from lab tests using SO_2 as impurity in CO_2 stream, were the rock porosity change and the chemical reactivity by ion migration with a relevant pH decrease in the effluent.
- Co-injection of CO_2 + artificial air (5% in volume $O_2 + N_2$) to study the effects of these impurities on the operation efficiency and reservoir capacity, revealed the necessity of 12.5% WHP increase to inject the mixture, and a relevant density decrease in gas trapping close to 8% (i.e. from 0.840 to 0.775 t/m^3).

Acknowledgements The experiences and results showed in this chapter form part of the projects "OXYCFB300" and "IMPACTS" funded by the European Energy Program for Recovery (EEPR), Seven Framework Program (FP7) and the Spanish Government through Foundation Ciudad de la Energía-CIUDEN F.S.P. Authors acknowledge the role of the funding entities, project partners and collaborators without which the project would not have been completed successfully.

This document reflects only the authors' view and that European Commission or Spanish Government are not liable for any use that may be made of the information contained therein.

Glossary

BHP Bottom hole pressure
BHT Bottom hole temperature
CCS Carbon capture and geological storage
DTS Distributed Temperature Sensing System

EEPR European Energy Program for Recovery
EPA Environmental Protection Agency
FEED Front End Engineering Design
FID Final investment decision
FP7 Framework Program 7th
HA Observation well
HI Injection well
LOT Leak off test
TDP Technology Development Plant
WAG Water alternating gas
WHP Well head pressure
WHT Well head temperature

References

1. Nazarian, B., Held, R., Hoier, L., & Ringrose, P. (2013). Reservoir management of CO_2 injection: Pressure control and capacity enhancement. *Energy Procedia, 37*, 4533–4543. Available online https://doi.org/10.1016/j.egypro.2013.06.360. Accessed March 17, 2020.
2. Burton, M., Kumar, N., & Bryant, S. (2009). CO_2 injectivity into brine aquifers: Why relative permeability matters as much as absolute permeability. *Energy Procedia, 1*, 3091–3098. GHGT-9. Available online https://core.ac.uk/download/pdf/82597741.pdf. Accessed March 17, 2020.
3. de Dios, J. C., Delgado, M. A., Marín, J. A., Salvador, I., Álvarez, I., Martinez, C., & Ramos, A. (2017). Hydraulic characterization of fractured carbonates for CO_2 geological storage: Experiences and lessons learned in Hontomín Technology Development Plant. *International Journal of Greenhouse Gas Control, 58C*, 185–200.
4. Rutqvist, J., Birkholzer, J., Cappa, F., & Tsang, C. (2007). Estimating maximum sustainable injection pressure during geological sequestration of CO_2 using coupled fluid flow and geomechanical fault-slip analysis. *Energy Conversion and Management, 48*(6), 1798–1807. Available online https://doi.org/10.1016/j.enconman.2007.01.021. Accessed March 17, 2020.
5. Chiaramonte, L., Zoback, M., Frieddman, J., & Stamp, V. (2007). Seal integrity and feasibility of CO_2 sequestration in the Teapot Dome EOR pilot: Geomechanical site characterization. *Springer Environmental Geology, 54*, 1667–1675. Available online https://link.springer.com/article/10.1007/s00254-007-0948-7. Accessed March 17, 2020.
6. *The Compostilla Project «OXYCFB300» Carbon capture and storage demonstration project. Knowledge sharing FEED report*. Global CCS Institute publications. Available online https://hub.globalccsinstitute.com/sites/default/files/publications/137158/Compostilla-project-OXYCFB300-carbon-capture-storage-demonstration-project-knowledge-sharing-FEED-report.pdf
7. Gastine, M., Berenblyum, R., Czernichowski-Lauriol, I., de Dios, J. C., Audigane, P., Hladik, V., et al. Enabling onshore CO_2 storage in Europe: Fostering international cooperation around pilot and test sites. *Energy Procedia, 114*, 5905–5915.
8. *Hontomín Technology Development Plant*. Available online https://www.enos-project.eu/sites/operational-storage-field-site/hontomin/
9. Bruant, R., Guswa, A., Celia, M., & Peters, C. (2002). *Safe storage of CO_2in deep saline aquifers*. American Chemical Society. Environmental Science and Technology. Available online https://pubs.acs.org/doi/pdf/10.1021/es0223325. Accessed March 19, 2020.
10. Vilarrasa, V., Carrera, J., & Olivella, S. (2013). Hydromechanical characterization of CO_2 injection sites. *International Journal of Greenhouse Gas Control, 19*, 665–677. Available online

https://www.sciencedirect.com/science/article/abs/pii/S1750583612002794. Accessed March 19, 2020.

11. *United States Environmental Protection Agency geologic CO_2sequestration technology and cost analysis*. Available online https://www.epa.gov/sites/production/files/2015-07/documents/support_uic_co2_technologyandcostanalysis.pdf. Accessed March 19, 2020.
12. Rutqvist, J., Bilkholzer, J., & Tsang, C. (2008). Coupled reservoir–geomechanical analysis of the potential for tensile and shear failure associated with CO_2 injection in multilayered reservoir–caprock systems. *International Journal of Rock Mechanics and Mining Science, 45*(2), 132–143. Available online https://doi.org/10.1016/j.ijrmms.2007.04.006. Accessed March 19, 2020.
13. Yang, S. O., Hamilton, S., Nixon, R., & de Silva, R. (2015). Prevention of hydrate formation in wells injecting CO_2 into the saline aquifer. *Society of Petroleum Engineers, 30*(1). Available online https://doi.org/10.2118/173894-PA. Accessed April 14, 2020.
14. Shindo, Y., Christer Lund, P., Fujioka, Y., & Komiyama, H. (1993). Kinetics and mechanism of the formation of CO_2 hydrate. *International Journal of Chemical Kinetics*. Available online https://doi.org/10.1002/kin.550250908. Accessed April 15, 2020.
15. Newton, L. E., & MacClay, R. A. (1977). Corrosion and operational problems, CO_2 project, Sacroc unit. In *Society of Petroleum Engineers. SPE Permian Basin Oil and Gas Recovery Conference*, Midland, TX, March 10–11, 1977. Available online https://doi.org/10.2118/6391-MS. Accessed April 20, 2020.
16. Ghahfarokhi, R. B., Pennell, S., Matson, M., & Linroth, M. (2016). *Overview of CO_2injection and WAG sensitivity in SACROC*. Society of Petroleum Engineers. Available online https://www.spe.org/en/jpt/jpt-article-detail/?art=2999. Accessed April 20, 2020.
17. de Dios, J. C., Le Gallo, Y., & Marín, J. A. (2019). Innovative CO_2 injection strategies in carbonates and advanced modeling for numerical investigation. *Fluids, 4*(1), 52. Available online https://doi.org/10.3390/fluids4010052
18. Oldenburg, C. M. (2006). Joule-Thomson cooling due to CO_2 injection into natural gas reservoirs. *Energy Conversion and Management, 48*(6). Available online https://doi.org/10.1016/j.enconman.2007.01.010. Accessed April 8, 2020.
19. Núñez López, V., Muñoz Torres, J., & Zeidouni, M. (2014). Temperature monitoring using distributed temperature sensing (DTS) technology. *Energy Procedia, 63*. 3984–3991. Available online https://doi.org/10.1016/j.egypro.2014.11.428. Accessed April 21, 2020.
20. Porter, R., Fairweather, M., Pourkashanian, M., & Woolley, R. (2015). The range and level of impurities in CO_2 streams from different carbon capture sources. *International Journal of Greenhouse Gas Control, 36*, 161–174. Available online https://doi.org/10.1016/j.ijggc.2015.02.016. Accessed April 22, 2020.
21. IMPACTS Project website. Available online https://www.sintef.no/projectweb/impacts/the-project/
22. de Dios, J. C., Delgado, M. A., Marín, J. A., Martinez, C., Ramos, A., Salvador, I., & Valle, L. (2016). Short-term effects of impurities in the CO_2 stream injected into fractured carbonates. *International Journal of Greenhouse Gas Control, 54*, 727–736. Available online https://doi.org/10.1016/j.ijggc.2016.08.032

Modeling Aspects of CO_2 Injection in a Network of Fractures

Srikanta Mishra, Samin Raziperchikolaee, and Yann Le Gallo

Abstract This chapter provides an overview of analytical and numerical modeling approaches for evaluating the effects of CO_2 injection into a network of fractures. The system of interest consists of two components—a number of potentially connected high-permeability but low porosity fractures embedded in a low-permeability but higher porosity matrix. The concept of injectivity index, based on analytical solutions to single-phase flow equations in an equivalent continuum, is first explained followed by field applications. The relationship between injectivity index and permeability is also explored based on field data and numerical simulations. Next, a hierarchy of numerical modeling approaches is described ranging from equivalent single continuum, dual porosity (flow only in idealized fractures), dual permeability (flow in fractures and matrix), and discrete fracture networks (flow in a complex fracture network and connected matrix). A case study of CO_2 injection into a depleted oil field in the Appalachian Basin, USA, is presented that involves the first three approaches referenced above, followed by a case study of modeling of CO_2 injection into a saline aquifer in Hontomin, Spain, using the discrete fracture network approach.

Keywords Carbon sequestration · Enhanced oil recovery · Deep saline aquifers · Depleted oil fields · Natural fractures · Injectivity index · Dual porosity · Dual permeability · Discrete fracture · Single continuum · Dual continuum · Hontomín pilot · Appalachian basin

S. Mishra (✉) · S. Raziperchikolaee
Battelle Memorial Institute, Columbus, OH, USA
e-mail: mishras@battelle.org

Y. Le Gallo
GEOGREEN, Rueil Malmaison, France

J. C. de Dios et al. (eds.), *CO_2 Injection in the Network of Carbonate Fractures*, Petroleum Engineering, https://doi.org/10.1007/978-3-030-62986-1_6

1 Purpose of Modeling

This chapter deals with the modeling process for simulating CO_2 injection and associated storage in saline aquifers, as well as the additional aspect of oil production in depleted oil fields—with a focus on systems characterized by a network of fractures. Typically, this modeling process entails two phases. The first phase, geologic framework modeling, integrates all pertinent geological and geophysical data (from logs, cores and seismic surveys) about reservoir structure, geometry, natural fracture network, rock types and property distributions (porosity, permeability, water saturation) into a 3-D distributed grid-based static earth model (SEM). The second phase, dynamic reservoir modeling, uses the SEM as a platform to simulate the movement of CO_2 and brine (and oil and gas as appropriate) and pressure change within the reservoir during CO_2 injection and brine/oil production.

These modeling studies support several goals, i.e.,

- Geologic System representation—data integration (e.g., integration of all reservoir characterization data into a geologic framework)
- Scientific—coupled process understanding (e.g., how does CO_2 move through the formation and interact with rock/oil/brine)
- Calibration—history matching (e.g., update description of subsurface by comparing model predictions to observations)
- Engineering—system design (e.g., how many wells are needed to meet injection targets and optimize CO_2 storage and oil recovery).

As is well known, for the purpose of pure CO_2 storage in the subsurface, deep saline formations (either sandstone, carbonate, or a stacked system) are considered primary candidates due to their worldwide presence and sizeable storage capacity [4, 20, 51]. CO_2 can also be stored in depleted oil reservoirs during or after the enhanced oil recovery EOR process. As a result, CO_2 injection can also be beneficial not only to increase production but also as a method to reduce the amount of greenhouse gases in the atmosphere [5, 45, 52].

The presence of natural fractures is common in carbonate reservoirs affecting EOR recovery as well as CO_2 storage [1, 61]. When gas injection method is considered to increase oil recovery, an early injectant gas breakthrough could affect the outcome of EOR plan. Injectant fluid could bypass the oil in the matrix due to higher permeability of fractures and mixing of oil and gas could be affected. Natural fractures and faults could also affect different aspects of CO_2 storage into saline aquifers. Presence of natural fracture networks could affect the practical storage capacity of aquifer [51], injection pressure uncertainty [6], leakage from the reservoirs by fault and fracture network activation due to stress changes (Vilarrasa et al. 2014) [18], and mineral reactions induced by CO_2-enriched water [44].

The practice of CO_2 injection for enhancing oil recovery in depleted oil fields has a long history. As such, different methods ranging from simplified to detailed numerical approaches, have been developed to address the challenges of CO_2 injection, evaluate efficiency of CO_2-EOR and associated storage, and enable an optimized EOR plan in carbonate reservoirs [45]. Numerical, analytical, and simplified screening tools have been used to evaluate the effect of different parameters, including injection strategies, operational parameters, rock and fluid properties on CO_2-EOR and storage performance [3, 19, 24, 25, 30].

In this chapter, two main approaches typically used to evaluate fluid flow behavior in a fracture network will be discussed. First, analytical approaches will be discussed. Then, numerical approaches, including both continuum and discrete modeling approaches, will be discussed. The objective of this chapter would be an overview of different fracture network modeling approaches by considering the distinct aspects of different approaches. The application of the different approaches for modeling CO_2-EOR and CO_2-storage into fractured carbonate reservoir will also be discussed.

2 Analytical Approaches

In this section, we discuss the role of simple lumped parameter models based on closed-form analytical expressions of reservoir behavior. The objective is to present simple equations that define the injectivity index which can relate injection rate to the corresponding pressure buildup via basic formation characteristics such as the permeability-thickness product. These models can serve as rapid assessment tools for predicting formation response to injection, or quick-look reservoir analyses tools for interpreting field data.

2.1 *Concept of Injectivity Index*

Injectivity index (or its complement, productivity index) is a commonly used concept in petroleum reservoir engineering to evaluate the capability of a well to inject fluids into (or produce fluids from) a porous and permeable formation [56]. Injectivity index can be a simple and useful metric for comparing the performance of two different reservoirs, or a given reservoir before and after a key event, such as injection of a fluid at a significantly different rate, or a well workover operation. It is defined as the ratio of the injection rate divided by the pressure difference between formation pressure and bottom-hole pressure:

$$J = \frac{q}{(P_i - P_{BH})} \tag{1}$$

where J is injectivity index, q is injection rate, P_i is reference formation pressure and P_{BH} is bottom-hole pressure. During a typical injection event, the injection rate, q, is maintained at a relatively constant rate, and the bottom-hole pressure rapidly increases to some equilibrium value, P_{BH}, after which it changes slowly.

The difference between this quasi-equilibrium value and some reference formation pressure, P_i (i.e., stable pressure prior to injection), is the denominator in Eq. (1).

Based on the theory of well-test analysis in gas reservoirs [31], we can write an expression for the injectivity index of the transient period during CO_2 injection as:

$$J = \frac{q_{SC}}{(P_i - P_{BH})} = \left(\frac{kh}{1422T}\right)\left(\frac{2P_i}{\mu_i z_i}\right)\left(\frac{1}{\ln\left({r_D}/{r_w}\right)}\right) \tag{2}$$

where k is permeability, h is formation thickness, T is reservoir temperature, P_{i} is initial reservoir pressure, μ_{i} is initial viscosity at P_{i}, z_{i} is gas deviation factor at P_{i}, q_{SC} is the CO_2 injection rate (at standard conditions) and r_{w} is the wellbore radius. The only time-dependent variable in this expression is the transient drainage radius r_D, which is proportional to the logarithm of (hydraulic diffusivity $\times$ time), and as such, changes slowly. In fact, experiences from field-scale CO_2 injection projects suggest that the injectivity index quickly reaches a quasi-stable value [35].

Similarly, for a closed reservoir under boundary dominated (pseudo-steady-state) flow, the injectivity index can be written as:

$$J = \frac{q_{SC}}{\left(\overline{P} - P_{BH}\right)} = \left(\frac{kh}{1422T}\right)\left(\frac{2P_i}{\mu_i z_i}\right)\left(\frac{1}{\ln\left(0.472{r_e}/{r_w}\right)}\right) \tag{3}$$

where r_{e} is the external radius of the closed system and $\overline{P}$ is the average reservoir pressure. Since the outer boundary is at a fixed distance, this equation suggests that the injectivity index should also reach a constant value. Note that for the pseudo-steady-state period, the injectivity index cannot be directly calculated as the average reservoir pressure is unknown. However, we can use the following identity:

$$\frac{P_i - P_{BH}}{q_{SC}} = \frac{P_i - \overline{P}}{q_{SC}} + \frac{\overline{P} - P_{BH}}{q_{SC}} = \left(\frac{Q}{q_{SC}}\right)\left(\frac{1}{V_p c_t}\right) + \frac{1}{J} \tag{4}$$

where the first term on the right follows from simple material balance considerations and the second term is the definition of the injectivity index. This suggests that injection well pressure build-up normalized by the injection rate, when plotted against the ratio of cumulative injection to injection rate (i.e., material balance time), should yield a straight line with slope inversely proportional to the pore volume (V_p) times total fluid compressibility (c_t), and intercept equal to the reciprocal of the stable injectivity index. This is generally referred to as a flowing material balance plot [42].

Mishra et al. [36] have observed that the injectivity index appears to correlate with the permeability thickness product, at least based on Eqs. (2) and (3). Figure 1 shows a cross-plot of these variables, taken from field as well as numerical experiments of CO_2 injection into both saline aquifers and depleted oil fields. A clear trend is evident from

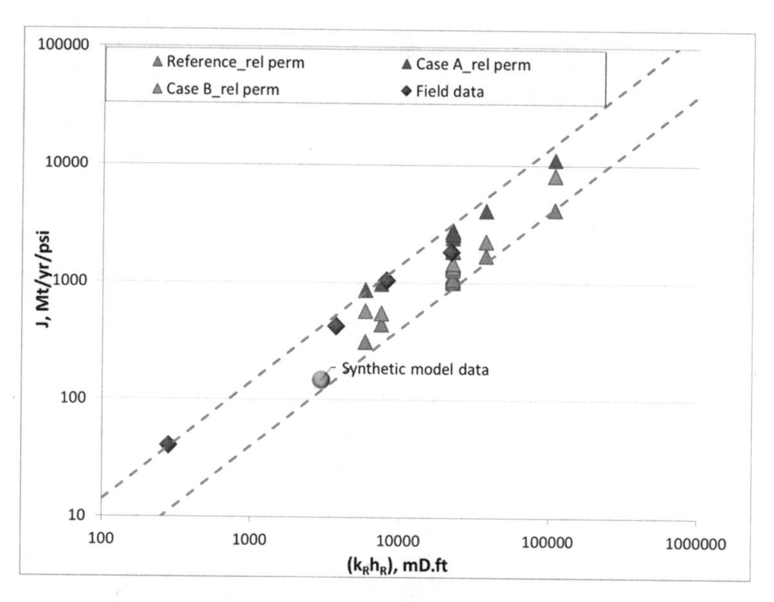

Fig. 1 Correlation of injectivity in different geologic settings with permeability based on field cases and simulated data (after [36])

this log-log plot, across multiple decades of variation in the permeability-thickness product. The scatter shown in the data is attributable to known differences in the characterization of relative permeability relationships (which is not considered in the above equations). The trend is well described by $J \sim 0.04\ kh$ at the lower bound and $J \sim 0.14\ kh$ at the upper bound—with $J \sim 0.1\ kh$ appearing to be adequate for scoping calculations.

2.2 *Applications*

Mishra et al. [35] presented an application of this concept for CO_2 injection into a saline aquifer. The injection zone of interest the vuggy dolomite Copper Ridge formation in the US Appalachian basin. Approximately 27,000 metric tons (MT) of supercritical CO_2 was injected into the Copper Ridge dolomite formation in well AEP-1 with pressure monitoring undertaken in MW-2, located ~2200 ft away. Injection rate and pressure history are shown in Fig. 2 for the Copper Ridge wells. It should be noted that for the Copper Ridge formation, significant pressure fluctuations were observed in the injection well pressure data after the September 2010 workover event, rendering this data unreliable and unusable.

The first step in the injectivity index calculations was to simplify the rate history shown in Fig. 2 by aggregating all injection (or shut-in) events less than 1000 min into the previous shut-in (or injection) event. For AEP-1, this resulted in 21 injection

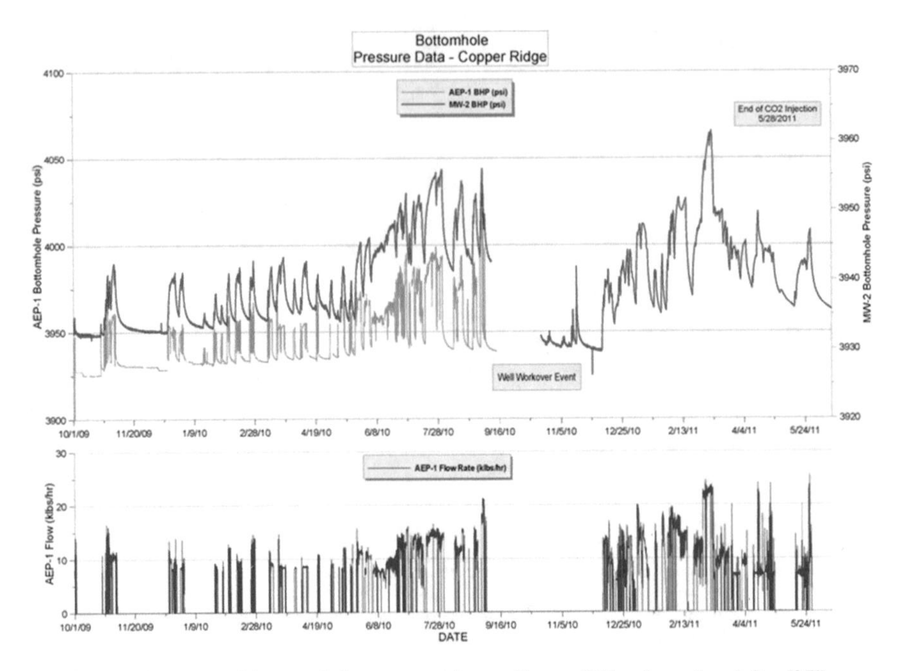

Fig. 2 Injection rate and bottom-hole pressure history, Copper Ridge formation (after [35])

events ranging from 2628 (~1.8 days) to 41,406 min (~29 days). The quasi-steady final pressure for each of these events was noted, along with the reference (pre-injection) pressure for each formation. Equation (1) was then used to calculate the injectivity index for each injection event.

Figure 3 shows the variability of injectivity index over time for the Copper Ridge formation. Note that no calculations could be performed for the period beyond t = 50,000 min because of the AEP-1 pressure gauge malfunction as mentioned earlier. The mean injectivity index was found to be 1800 MT/yr/psi, with a 25th percentile value of 1500 MT/yr/psi and a 75th percentile value of 2050 MT/yr/psi—which is a reasonably consistent range, given the variability in the injection rate (as shown in Fig. 2).

One can interpret these values as follows: on the average, the formation can accept ~1800 MT/yr for each psi of pressure buildup from pre-injection hydrostatic conditions. Also, the formation permeability thickness, based on well-test analyses, were found to be ~23,000 mD ft. These values are plotted in Fig. 1 as the highest filled red diamond, in line with the overall trend between J and kh.

Next, we discuss the application of the flowing material balance plots for calculating injectivity index in two depleted oil fields (i.e., closed reservoirs) undergoing CO_2 injection in the Northern Michigan Pinnacle Reef Trend [37]. Figure 4 shows injectivity index calculations for two reefs in Michigan, Reef A and Reef B, using flowing material balance plots. The red lines represent the straight-line fits performed to the data indicated as blue circles. Based on the intercepts of the linear trendlines

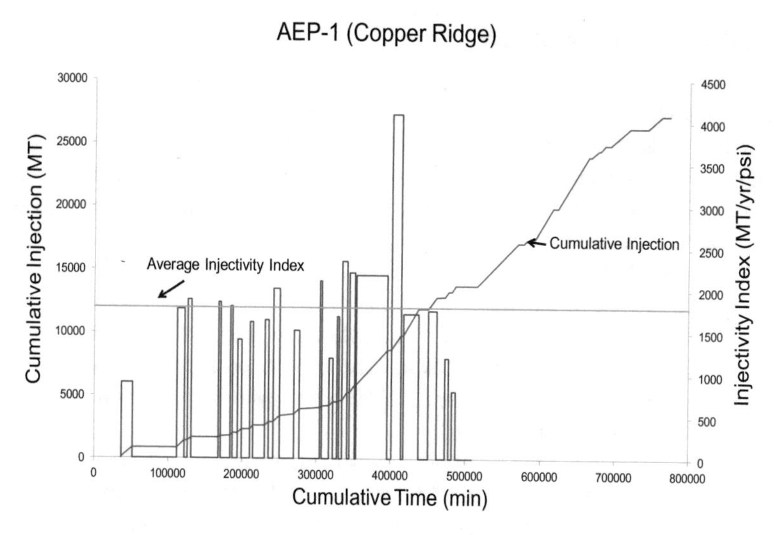

Fig. 3 Calculated injectivity index at AEP-1, Copper Ridge formation (after [35])

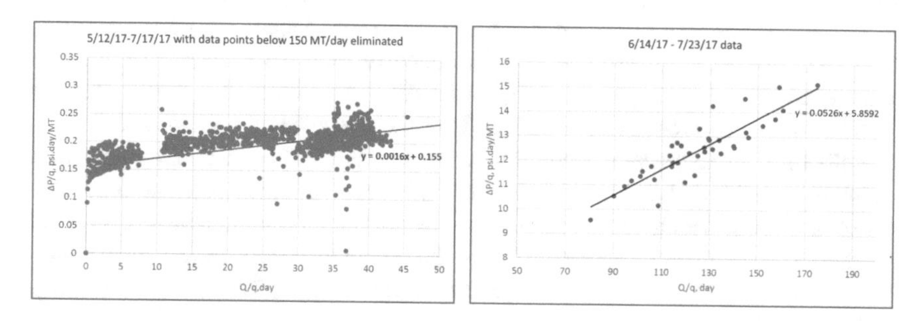

Fig. 4 Flowing material balance plots for Reef A (left); and Reef B (right); after Mishra et al. [37]

in these plots, the injectivity index values are calculated to be 2807 metric tons (MT)/year/psi for Reef A (left panel) and 62 MT/year/psi for Reef B (right panel).

While well-test derived permeability-thickness values are not available for these reefs, the "global" correlation in Fig. 1 can also be used to estimate the *kh* product from estimated injectivity index. From the trend line, the inferred permeability-thickness product for C-19 is ~25,000 mD-ft, while that for C-16 is ~500 mD-ft, and indicates that that C-19 reservoir performance is much more conducive to CO_2 injection than C-16.

3 Numerical Approaches

One of the geological characteristics that could significantly affect fluid flow behavior of the reservoir is the presence of natural fractures. We use the term natural fracture to designate pre-existing weakness planes, including joints, fissures, and fractures, regardless of the mechanism (e.g., shear, tensile, or mixed modes) that generated them. Numerical modeling from micro- to macroscales (field scale) shows the important role of natural fractures on the fluid flow behavior of reservoirs. In micro-scale modeling, geological, mechanical, and geometrical properties of a single or limited number of fractures are the objects of investigation [48]. Micro-scale simulation can be used to characterize flow transport directly on pore space images obtained from the scanning experiments generating 3D pore space images at different spatial resolutions of a fractured sample [14].

In the macro-scale modeling approaches, on the other hand, fracture network properties (density, orientations, and interactions) are emphasized to investigate their effects on fluid flow behavior. Different macro-scale modeling approaches were introduced and applied to model fractured media and accurately consider the effect of fracture networks on fluid flow behavior. Occurrence of multiphase flow combined with implicit modeling of fractures is a major challenge since the non-linearity of the system should be included due to presence of multiple phases with different compressibility, relative permeability, capillarity, and wettability. We discuss the main methods that were extensively used for the presence of fracture networks at the field scale (macro-scale) including equivalent single medium continuum, dual continuum model, and discrete fracture network approaches [9, 60]. The schematic of these approaches is shown in Fig. 5. The above-mentioned approaches are different in terms of the geometric representation of the fracture network as well as the interaction between matrix and fracture medium. Depending on the model complexity, computational time to investigate fluid flow behavior in a fractured medium through numerical simulations would also be different.

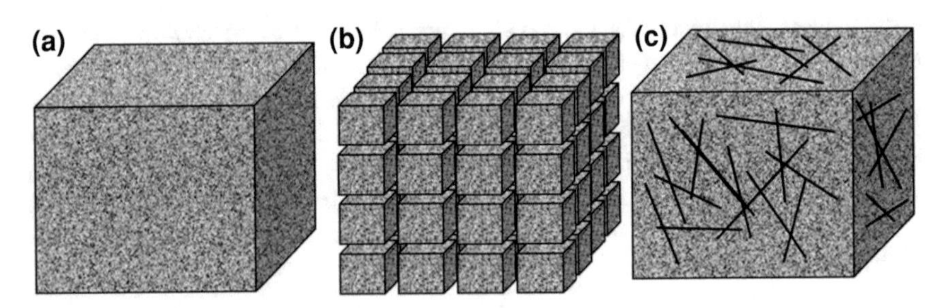

Fig. 5 Schematic of different modeling approaches: **a** single continuum, **b** dual continuum, **c** fracture network models (after [59])

3.1 Single Equivalent Continuum

Single equivalent continuum approach is considered as the simplest conceptual model for fractured porous media. In this model, the fracture properties are incorporated in the model by changing the porous media (i.e., matrix) permeability. The effective permeability of fractured medium depends on fracture network properties including distribution, orientation, density, size, and shape of fractures [10, 33, 60]. Different upscaling approaches can be used to estimate the effective permeability of fractured system as a function of above-mentioned fracture network properties as well as the matrix geological properties. If multiple fracture sets are present, the effective permeability tensor can be computed by summing the permeability tensors for each fracture set [39]. The effective permeability of the fractured medium can then be used in the continuity equation (mass conservation equation) and solved by an appropriate numerical approach to understand the effect of fractures on fluid flow behavior of reservoirs. Using a single equivalent continuum approach, the computational cost of modeling would be the lowest between the three methods. On the other hand, it is the most simplified model as no explicit terms related to the fracture network will be added to the continuity equation.

Single continuum modeling approach was used to evaluate CO_2-EOR and storage in fractured reservoirs [41, 49]. Different methods (from simpler analytical to semi-analytical methods) were used to estimate the effective permeability using the single continuum method [60]. Effective permeability method was used to evaluate CO_2-EOR and storage performance of the fractured Tensleep reservoir at Teapot Dome (Wyoming, USA) using a field-scale numerical simulation. The effective permeability, depending on fracture density and matrix permeability, was used as a parameter for the history match of CO_2-EOR process [41].

Field injectivity tests and appropriate well log data can also provide an estimate of effective permeability of a fracture network. Over 700 observations of natural fractures on acquired image logs collected at multiple well locations ranging in depth from 730 to 3900 m in the Knox Group interval on the western flank of Appalachian Basin (Ohio, USA) shows the presence of natural fractures specifically in Copper Ridge Dolomite aquifer [50]. The nuclear magnetic resonance (NMR) log was used to measure permeability of fractured zone, presenting the effective permeability of the formation. The nuclear magnetic resonance (NMR) log shows that the permeable zones in the Copper Ridge dolomite are in reasonable agreement with the fractured zones predicted by the image log. The zone-by-zone permeability measurements provided by the NMR log was imported to model layers of fractured zones in the carbonate reservoir (Fig. 6). Then, numerical simulation was performed to estimate the total injected mass of CO_2 associated with layers of natural fractures (presented by effective permeability) in the Copper Ridge dolomite aquifer after 30 years of injection.

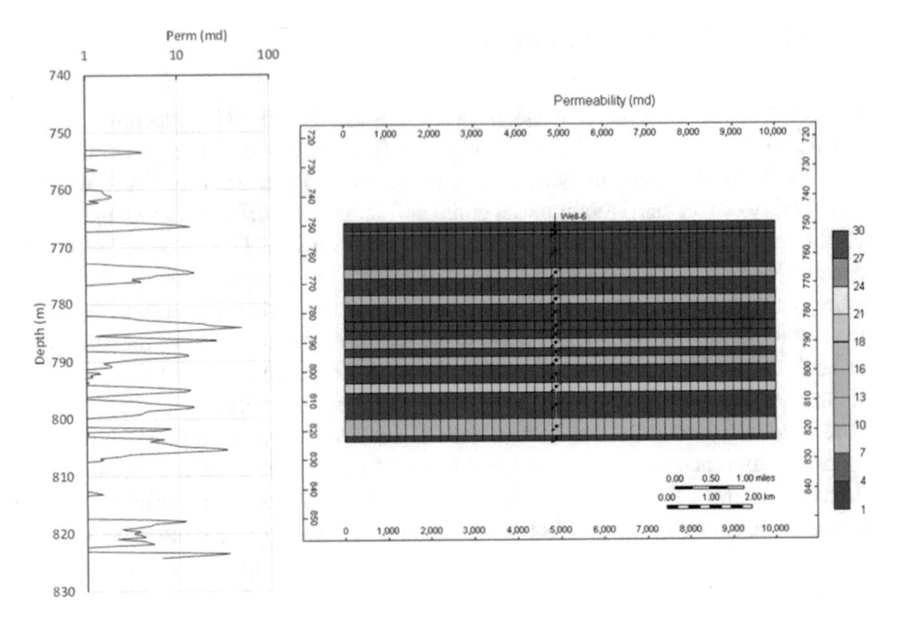

Fig. 6 Zones with effective permeability assigned by NMR log in the Upper Copper Ridge dolomite. NMR logs permeability (left) and modeled effective permeability in simulation (right) (after [50]). X and y-axis scales indicate distance (ft)

3.2 Dual Continuum Approach

In dual continuum approach, two media (fractured and matrix medium) are interacting with each other to represent the fluid flow behavior of the fractured medium [59]. As a result, two continuity equations (mass conservation equations) should be solved for matrix and fractured media with a transfer term describing the interaction between the two media [28]. The transfer term depends on interaction parameters including fracture matrix interface area [22].

Simulation of fractured reservoirs using dual continuum (e.g., the dual-porosity, dual-permeability) approach involves discretization of the solution domain into two continua, called the matrix and the fracture [27, 59]. In this model, rectilinear prisms of the rock matrix are separated by continuum of fractures. A schematic of this representation is shown in Fig. 5b. The dual porosity/permeability model allows each simulator block to have up to two porosity systems, one called its matrix porosity and the other called its fracture porosity. Each system can have its own porosity value and its own permeability. The matrix and fracture domains are linked to each other through a transfer term that connects each fracture cell to its corresponding matrix cell in a grid block. The dual-permeability model is an enhancement to the standard dual-porosity model. In this model, the communication between the matrixes (the intergranular void space which is also referred to as the primary porosity) is not assumed to be negligible. The differences in terms of interblock communication

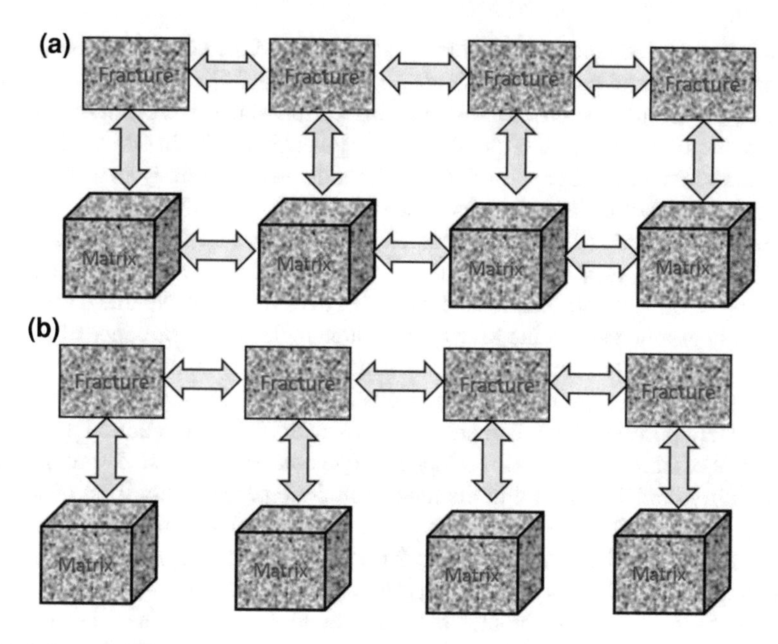

Fig. 7 Schematic diagrams of connectivity for **a** the dual permeability model and the **b** dual porosity models

for the standard dual porosity model and the dual-permeability model are shown in Fig. 7. Although a dual-porosity/permeability model can be used to represent fractured media, it is still a simple representation of a complex natural-fracture system since the fractures are assumed to be in a regular spacing.

Dual continuum model is the most common method to evaluate different aspects of fluid transport in fractured carbonate reservoirs. March et al. [34] discussed the role of the transfer function between matrix and fracture to accurately capture CO_2 transfer into matrix for CO_2 storage in a fractured reservoir using dual porosity model. In another study, a dual permeability model was built for a fractured carbonate reservoir in Qatar using the upscaling of a fracture network system to evaluate the effectiveness of CO_2 injection and storage [2]. The fractured system is represented using discrete fracture network (DFN) models by considering fracture observations (length, orientation, sets) in the outcrop and upscaled to build a dual permeability model. The EOR efficiency was then evaluated as a function of heterogeneity for the carbonate reservoir. The simulations were performed to investigate the sensitivities of different parameters (fracture intensity, wettability fracture geometry, trapping) affecting oil recovery. The simulation results show that the fracture characteristics are the key uncertainty for recovery predictions during CO_2-EOR.

Dual continuum models can be combined with a fracture activation model to study the permeability enhancement and its effect on reservoir performance. Different constitutive models were developed to relate the fractures activation to the permeability enhancement including a constant permeability increase [54, 58] and stress-dependent permeability increase models [53, 55].

Next, we discuss the case study involving performance of CO_2-EOR and associated storage in a depleted oil field of the Appalachian basin in Ohio, USA, using dual continuum modeling approach (see [7], for details). The study area is located in the northern part of the Morrow Consolidated oil field (MCOF) (Fig. 8). MCOF oil field was discovered in 1959. It has an area of 16,000 acres. This field is a promising candidate for CO_2 EOR as a result of poor primary recovery efficiency (i.e., 26%). The Copper Ridge dolomite in MCOF is known to be fractured. Natural fracture data acquired in eleven wells in the MCOF study area indicates the presence of fractures in the study area. Core descriptions and image logs were available and used to identify fractures. The map shows that the wells with a high density of fractures in the northwestern portion of the field are near areas with high production in the MCOF. In the case of fractured reservoirs (Copper Ridge dolomite in the study area), a dual-permeability model was used to distinguish fracture permeability from intact rock (matrix) permeability. The permeability of each reservoir grid-block represents the permeability of the fractures and the rock matrix in this model. The communication between the matrix blocks is also not assumed to be negligible.

Table 1 summarize the parameters used to build the model. The relative permeability curves used for matrix media is shown in Fig. 9, while linear relative permeability curves are used for fractured media.

Fluid composition of MCOF was obtained from field data sampling and modeled by Peng-Robinson EOS [46] which led to the identification of pseudo components using a regression procedure to minimize an objective function that is the difference between EOS predicted and experimentally determined fluid properties. The lumped composition to build a fluid model is shown in Table 2.

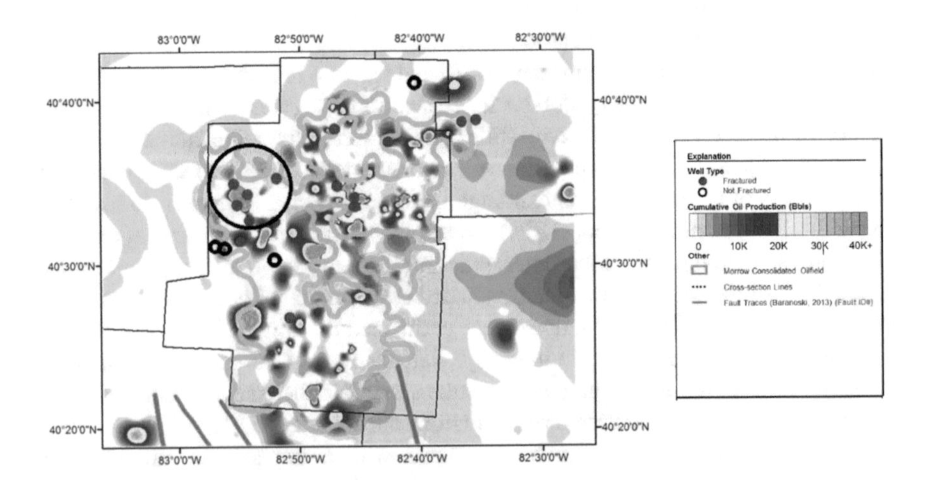

Fig. 8 Location of possible study area highlighted by black circle

Table 1 Key input parameters of the base case for MCOF

Parameter	Value
Initial pressure	1200 psi
Producer BHP	200 psi
Injector BHP	1200 psi
Porosity	0.04
Porosity—fracture	0.01
Formation top	3292 ft
Permeability fracture	50 mD
Permeability	1 mD
Model dimension	990 × 990 × 30 ft
Model grid numbers	11 × 11 × 10
Water saturation	0.3
Fracture spacing (vertical fractures)	30 ft

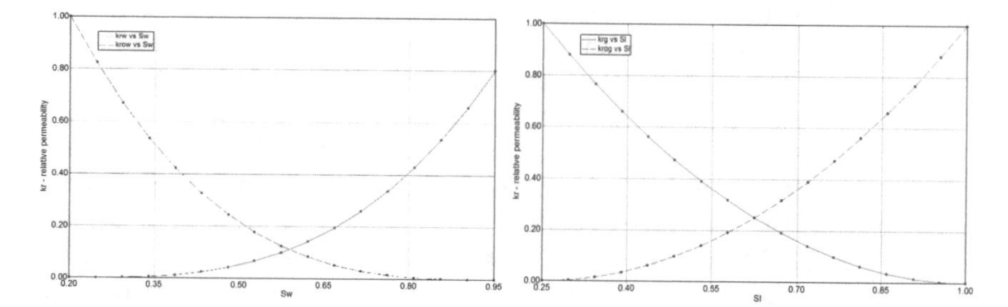

Fig. 9 Matrix relative permeability for oil-water (left) and gas-oil (right) for MCOF

Table 2 Lumped component model composition for MCOF sample

ID	Component	Mole (%)
N2	N_2	1.51
CO_2	CO_2	0.02
PC1	CH_4	21.36
PC2	C_2H_6 to C_3H_8	10.46
PC3	i-C_4H_{10} to C_6H_{14}	13.16
PC4	C_7H_{16} to $C_{17}H_{36}$	39.18
PC5	$C_{18}H_{38}$ to C_{36+}	14.31

In order to address the modeling objective, numerical models were constructed using an equation of state based reservoir simulator, CMG-GEM [11]. Key features of the numerical models are: (1) The rectangular Cartesian grids have uniform layer depth/thickness, (2) A single reservoir is sealed at the top of the formations, (3) A CO_2 injection well and production well are located at opposite corners of the model (taken as an element of symmetry of a 5 spot pattern), (4) Intact rock properties do not vary in the lateral/vertical direction. Heterogeneity stems from the presence of natural fractures, (5) The model represents a quarter of five-spot pattern, (6) A sealed outer model boundary is established without any aquifer drive. The CMG-GEM simulator models CO_2 behavior in an oil reservoir by solving an equation describing the thermodynamic equilibrium between gas, oil, and aqueous phases. The Peng-Robinson equation of state was used to model the phase behavior of the CO_2 and oil. CMG–GEM solves a discretized form of differential equations to step through time in the model.

A numerical scenario was performed to evaluate the significance of natural fractures on oil recovery and long-term CO_2 storage in a representative pattern where the injection well and production well are located in the corners of the model. Oil production increment was the primary parameter to be evaluated. In the model, a CO_2 injection rate was not specified as an input parameter but instead was calculated by the model. In this case, a maximum bottom-hole injection pressure (BHP) was specified, and the model determined the maximum injection rate that could be realized without exceeding the maximum BHP. CO_2 was injected into the wellbore six years after reservoir primary production. The oil increment was investigated after 6 months of the fill-up phase. The cumulative oil production and production rate during primary production and subsequent CO_2-EOR phase are shown in Fig. 10.

The fraction of CO_2 in the gas phase, oil saturation, pressure, and oil viscosity in different grid blocks of the reservoir were studied at the end of CO_2-EOR period for the fractured and not-fractured cases (Figs. 11 and 12). A significant amount of oil was swept by CO_2 at the end of the CO_2-EOR period. After 10 years of CO_2 injection, the amount of oil produced was significantly increased in the fractured scenario and the average oil saturation in the reservoir was reduced from initial oil saturation (0.7) to 0.33. Figure 11 shows the fraction of CO_2 in the gas phase, oil saturation, pressure, and oil viscosity for the case with natural fractures present. After 10 years of CO_2 injection, the amount of oil produced was 19,444 bbl in the case without natural fractures. The average oil saturation in the reservoir was reduced from initial oil saturation (0.7) to 0.58. The remaining oil in the reservoir was higher in comparison with the naturally fractured reservoir. The main reason for this was that less CO_2 was injected before reaching the specified BHP.

Several simulations were performed to history match the primary phase (before CO_2-EOR) oil production of the fractured model with an equivalent not-fractured model using an effective permeability. The purpose of this work was to study how EOR phases of fractured (using dual continuum approach) and matched intact models (using single continuum approach) differ from each other. In this case, we can better understand the role of fractures on oil recovery using CO_2 injection. All intact single continuum models were simulated using a permeability higher than the matrix permeability in the fractured case to achieve the primary oil production observed in the fractured case scenario. The higher matrix reservoir permeability was chosen for intact reservoir cases to estimate the effective permeability for the fractured reservoir. Figure 13 shows the result of history matching of the primary production. The fractured case using dual continuum modeling approach is shown by red line. Results

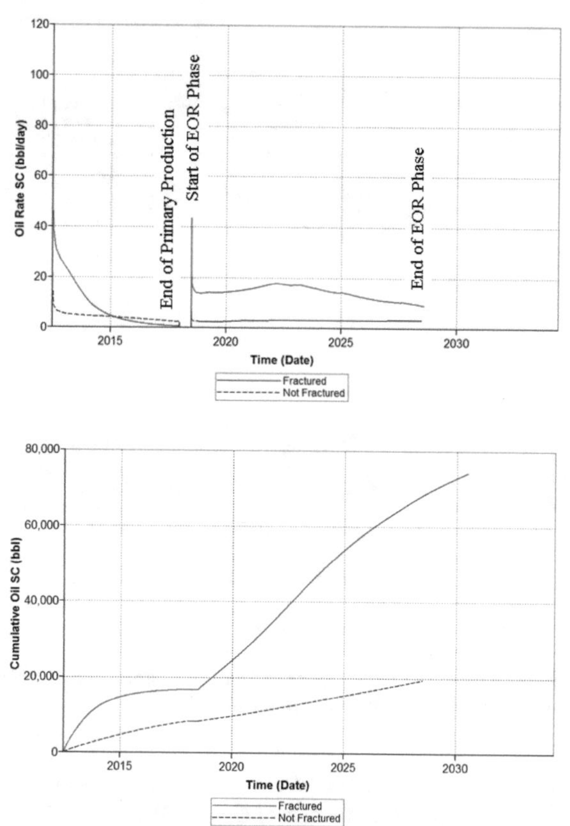

Fig. 10 Oil rate production and cumulative oil production for the case with and without fractures, includes historical data until 2017 and forecasts thereafter

of history matching show that the best permeability scenarios to match the primary production of fractured scenario are: (1) Kx = 20 mD, Ky = 20 mD, Kz = 5 mD; (2) Kx = 30 mD, Ky = 30 mD, Kz = 5 mD. Although matched cases produced a similar amount of oil at the end of primary production, both matched cases showed higher oil recovery after the EOR period in comparison to the fractured case modeled using dual continuum approach. This suggests that, EOR phase oil recovery is lower by presenting fractures using the dual continuum model compared to single continuum media matched case. This could be an artifact of the enhanced values assigned to permeability for the single continuum case, versus the assumed values for fracture and matrix permeabilities for the dual continuum case (as opposed to being derived by matching to field data). The simulation results shows the limitation of modeling multiphase flow in fractured media using single continuum media since relative permeability and capillary behavior of multiphase flow in fracture network is neglected in single continuum media.

Figure 14 captures the effect of these successive phases on cumulative oil production and net CO_2 stored. It also shows that oil recovered during the EOR phase is greater than that recovered during the first production phase, highlighting the effectiveness of CO_2 injection in sweeping the remaining oil out of the reservoir. CO_2

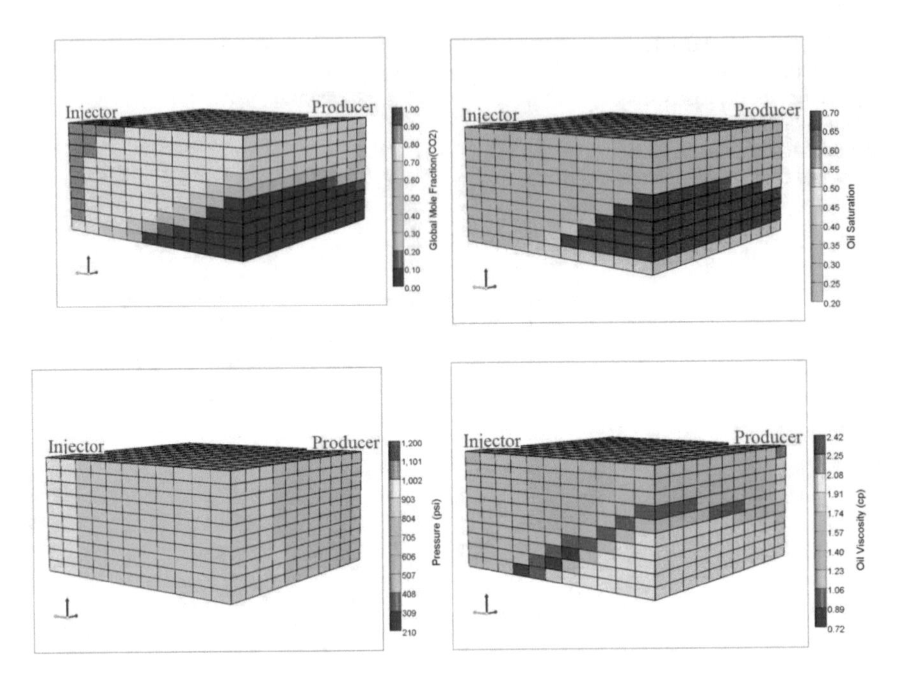

Fig. 11 CO_2 saturation, oil saturation, pressure, and oil viscosity at the end of CO_2-EOR when natural fractures are included in the model

storage occurs primarily during the reservoir re-pressurizing phase and the final CO_2 storage phase after the end of CO_2-EOR.

3.3 *Discrete Fracture Models*

In discrete fracture modeling, fracture networks are presented explicitly as discrete features, without employing upscaling methods, to represent fractured medium. A discrete fracture is a realization of the statistical model and is a 3-D representation of the network of fractures. The models based on explicit representation of fracture network would be complex with higher computational time by including realistic geometry of the fracture network [38]. If the matrix permeability is high enough and cannot be neglected, the fluid flow in the matrix medium should be taken into account to build a discrete fracture matrix model. Embedded discrete fracture model (EDFM) is one of the example of the discrete fracture models in which the coupling between matrix and fracture cell is taken into account (similar to dual continuum models) [32]. If the matrix permeability is very low or impermeable, a discrete fracture network (DFN) model can be used. In DFN model, the fluid is present only in the fracture network since the matrix media is assumed impermeable. Ngo et al. [40] modeled single phase solute transport within a complex and realistic DFN of the Bloemendaal

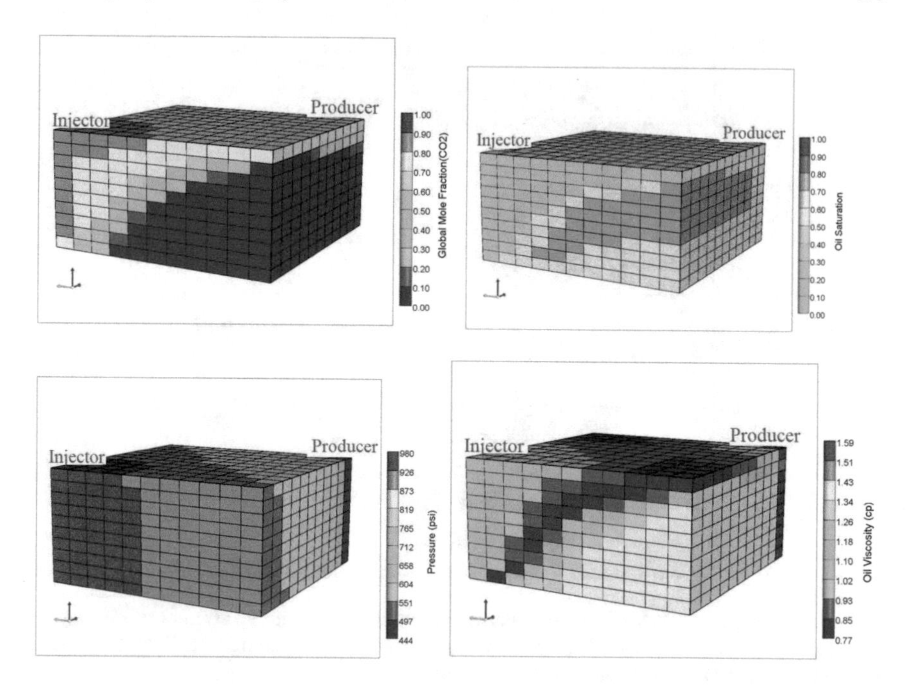

Fig. 12 CO_2 saturation, oil saturation, pressure, and oil viscosity at the end of CO_2-EOR when there are no natural fractures accounted for in the model

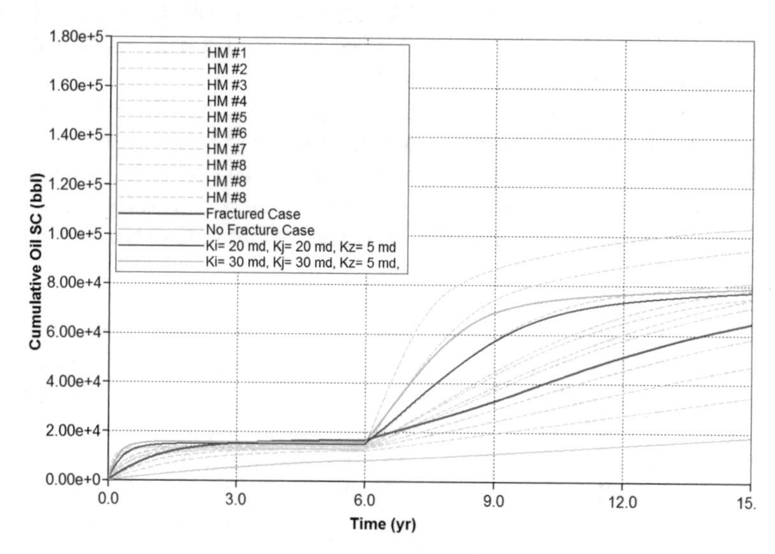

Fig. 13 Comparing oil recovery of fractured case with intact case having different permeabilities

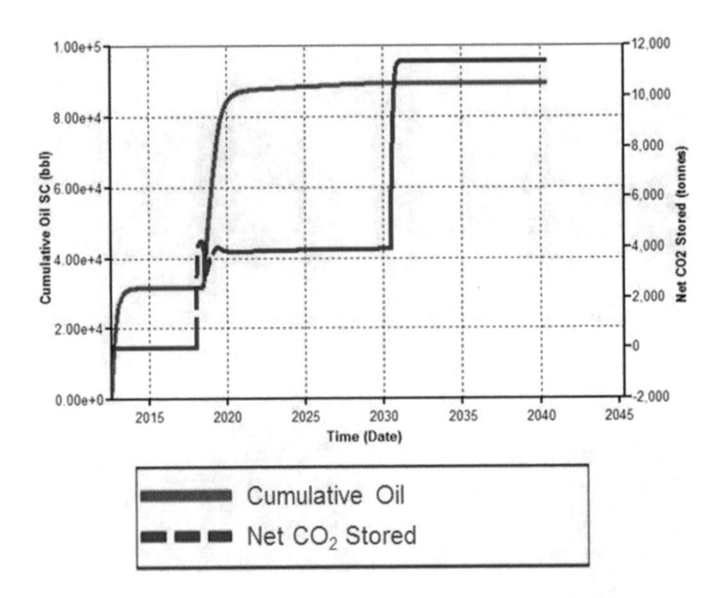

Fig. 14 The red curve (cumulative oil) shows that a total of around 90,000 bbl of oil may be recovered by the end of CO_2-EOR, while the blue curve (net CO_2 stored) suggests that around 11,000 tonnes of CO_2 may be stored in the reservoir

reservoir [57]. The network contains 3088 fractures and the mesh consists of about 1.1 million grid cells for a 3×3 km^2 region. The simulations showed the good connectivity within the fractured network (Fig. 15).

As numerous and important oil and gas reservoirs are carbonated, a large amount of work was carried out to improve their characterization and subsequent production. The corresponding workflows usually require DFN or EDFM to be matched with image logs which may then be matched with field or production data [12, 47]. The next step is the upscaling of the DFN to enable dual-porosity/dual-permeability modeling.

A similar workflow was implemented by Le Gallo and De Dios [21] on the Hontomìn site (Spain): the approach relies upon a EDFM construction from the identified fracture and calibration of the various parameters, such as permeability of each fracture set in all directions, aperture of each fracture set using well tests [26]. A specialized software, BF-FracaFlow [8] used an "automated *kh* calibration" method (detailed in [15]) which is based upon an analytical calibration of the EDFM parameters on the well test interpretations. The Hontomín storage reservoir comprises naturally fractured limestones and dolomites [17]. At Hontomìn, all wells are sub-vertical, thus limiting the intercepted fractures to the set of diffuse fractures and the fault zone above the storage formation that were identified during drilling. Most of the fractures identified in the wells are sub-horizontal and reflects the formation bedding. Two main sets of diffuse fractures were identified [21]:

- one with an approximate North-South (N-S) orientation (strike ~176 N);

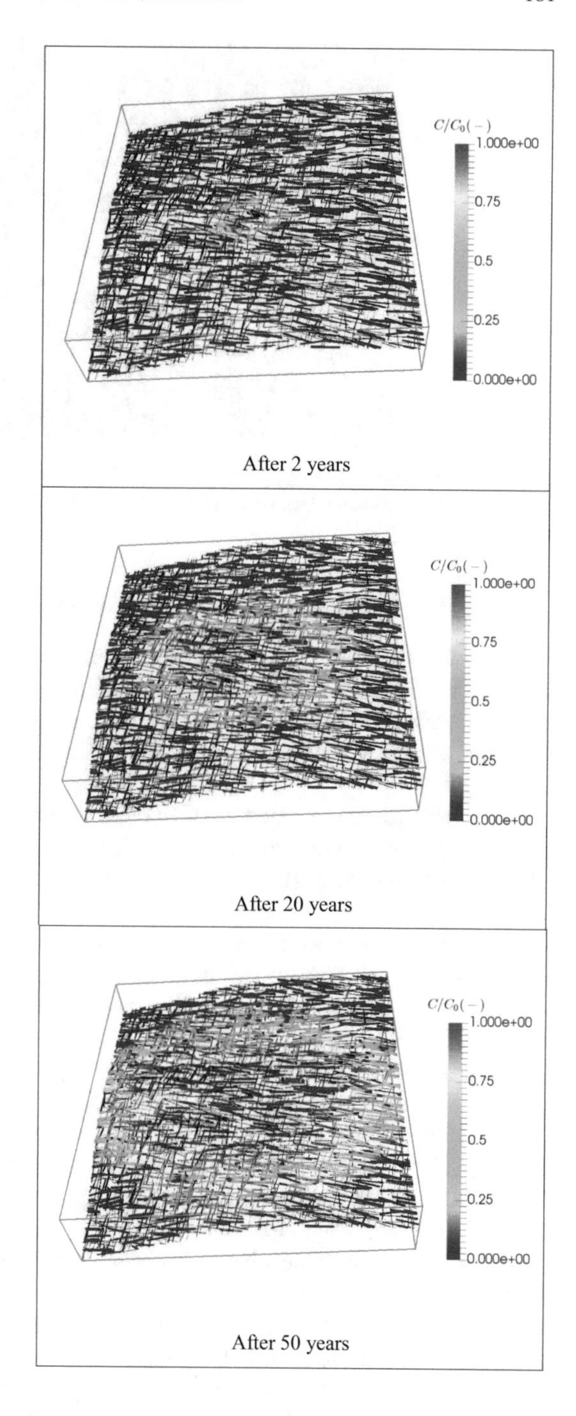

Fig. 15 Solute concentration distribution over time [40]

Fig. 16 Calibrated conductivity (mD m) of the E-W (blue) and in N-S (red) diffuse fracture networks in EDFM around injection (yellow) and observation (green) wells within the storage formation at Hontomin

- one with an approximate East-West (E-W) orientation (strike ~85 N).

The matrix properties were obtained from the storage geological model and the match of the interpreted flow capacity (*kh*) of the well test was performed on the fracture conductivities (Fig. 16) assuming a large number of realizations of the EDFM. The final results minimize the pressure differences with the interpreted well test results [29].

Once the hydraulic properties of fractures were calibrated on the well test interpretations [17], they were upscaled at the full-field grid size and equivalent properties of the fracture networks (Fig. 17) were computed and implemented in dual-medium continuum simulations at field scale.

The injection tests performed at Hontomìn [16] alternated CO_2 and brine injection periods. Figure 18 show the evolution of the bottom-hole pressure computed by the model and induced by the changes of injection conditions (flow rate, CO_2/brine ratio) using the history matched model over the single-phase brine injection periods [16].

The dual permeability modeling with CMG-GEM (CMG 2012) indicates that CO_2 only spread about 80 m away from the injection well (Fig. 19) during the CO_2-brine hydraulic injection tests (Fig. 18) and did not reach the observation well which is confirmed by field data [16]. The pressure impact extended beyond the CO_2 zone to about 400 m but the zone of high pressure (greater than 200 kPa) was restricted to the CO_2 extension (Fig. 19).

Discrete fracture models have also been used to evaluate the fluid flow behavior of fractured carbonate reservoirs [13, 23]. Panfili and Cominelli [43] used an EDFM to evaluate miscible gas injection in fractured carbonate reservoir in which corner point grid geometry for matrix are combined with an unstructured network for fractures. The modeling results were used to study gas breakthrough through a realistic presentation of fracture network. Figure 20 shows the mole fraction of gas in the fractures using EDFM and dual continuum model. The simulation results show different

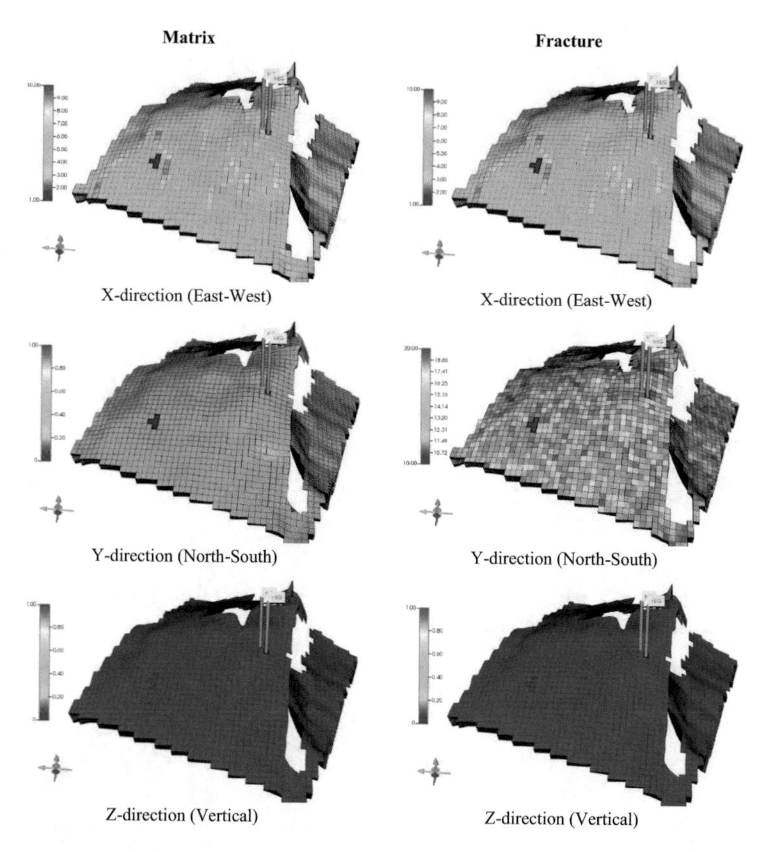

Fig. 17 Upscaled matrix and fracture permeabilities (mD) in the storage formation in the Hontomìn dual permeability model

pattern using different models. In comparison to a dual continuum model, EDFM presents fluid flow along fracture network in a realistic way by considering connectivity between fractures using explicit presentation of fracture geometry. Simulation results show a high gas saturation in a cluster of fractures connecting injector and producer wells using EDFM. As a result of that, an earlier break-through and higher gas volume production were estimated using EDFM approach.

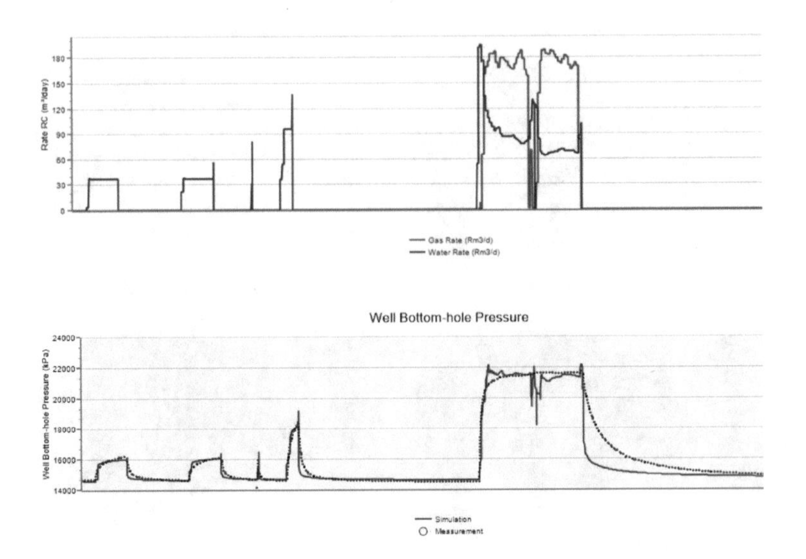

Fig. 18 Evolution of the bottom hole pressure at the injection well (bottom) induced by the brine and CO_2 injection tests at Hontomìn

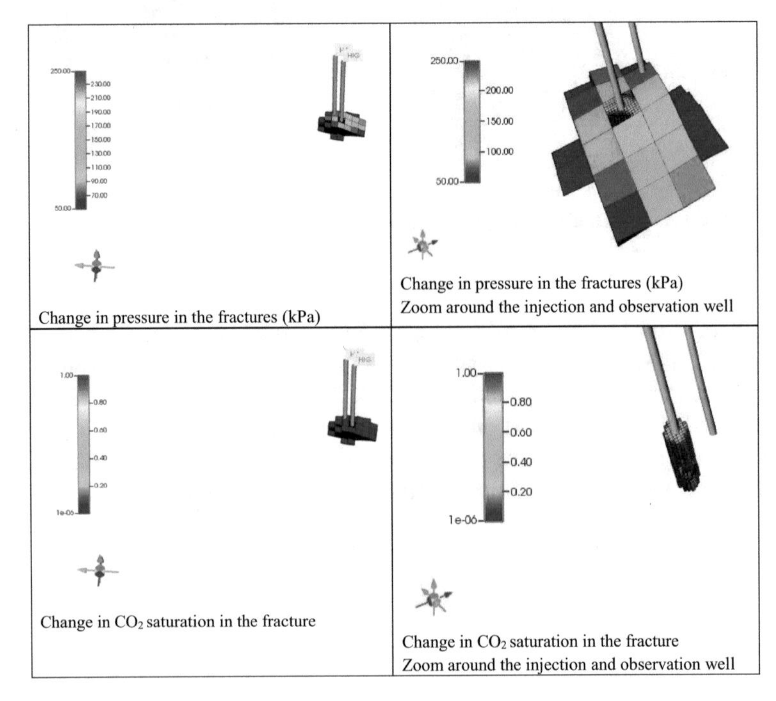

Fig. 19 Pressure and CO_2 saturation in the fracture within Hontomìn storage formation around the injection well at the end of CO_2-brine injection tests

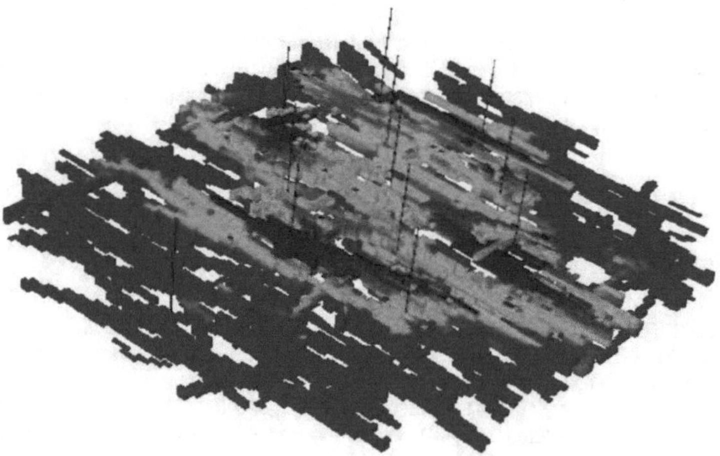

Fig. 20 CO_2 mole fraction at the end of simulations for the embedded fracture model (top) and dual continuum model (bottom)

4 Concluding Remarks

During past decades, different types of modeling approaches to present fracture network as well as the discretization methods were introduced. At the simplest level, lumped parameter concepts such as injectivity index based on analytical models can provide a quick-look representation of the system of interest. At a more detailed level, fracture network complexity in the reservoir can help determine the appropriate numerical modeling approach. A simpler less dense fracture network may justify a single continuum approach for modeling. On the other hand, a dense complex network of fracture requires dual continuum or an explicit presentation of fracture network. Upscaling of the fracture network is an important step for modeling the network implicitly. Multi-phase flow adds extra challenges for modeling the fractured medium. Including the multiphase flow phenomena in a dual continuum modeling approach is more developed.

Discrete fracture modeling approach can address simulation of naturally fractured systems for field scale reservoir studies by realistic presentation of fracture geometry. Challenges are still existed regarding incorporating different aspects of fracture properties (such as fracture geometry, density, and intersections) in a fracture network modeling approach to evaluate the fluid flow behavior during EOR including: fracture geometry presentation, host grid modification, and well completion modification automatically (using a graphical user interface). Multiphase flow modeling coupled with explicit representation of fracture network can also cause additional complexity and computational costs. Next generation high performance reservoir simulation can take benefit of a more detailed realistic presentation of fracture network using a discrete fracture model approach.

Acknowledgements The research described in this chapter was funded in parts by the United States Department of Energy National Energy Technology Laboratory (NETL) under award #DE-FC26-0NT42589, the Ohio Development Services Agency's Coal Development Office Grant D-15-08, and American Electric Power.

References

1. Agada, S., Geiger, S., & Doster, F. (2016). Wettability, hysteresis and fracture–matrix interaction during CO_2 EOR and storage in fractured carbonate reservoirs. *International Journal of Greenhouse Gas Control, 46,* 57–75.
2. Agada, S., & Geiger, S. (2014). Wettability, trapping and fracture-matrix interaction during WAG injection in fractured carbonate reservoirs. In *SPE Improved Oil Recovery Symposium.* Society of Petroleum Engineers.
3. Asghari, K., Dong, M., Shire, J., Coleridge, T. J., Nagrampa, J., & Grassick, J. (2007). Development of a correlation between performance of CO_2 flooding and the past performance of waterflooding in Weyburn oil field. *SPE Production & Operations, 22*(02), 260–264.
4. Bachu, S., Gunter, W., & Perkins, E. (1994). Aquifer disposal of CO_2: Hydrodynamic and mineral trapping. *Energy Conversion and Management, 35*(4), 269–279.
5. Bachu, S. (2004). Evaluation of CO_2 sequestration capacity in oil and gas reservoirs in the Western Canada Sedimentary Basin. In *Alberta Geological Survey, Alberta Energy and Utilities Board March* (pp. 1–77).
6. Bao, J., Hou, Z., Fang, Y., Ren, H., & Lin, G. (2015). Uncertainty quantification for evaluating the impacts of fracture zone on pressure build-up and ground surface uplift during geological CO_2 sequestration. *Greenhouse Gases: Science and Technology, 5*(3), 254–267.
7. Battelle. (2017). *CO_2utilization for enhanced oil recovery and geological storage in Ohio* (Reservoir characterization topical report). Battelle.
8. Beicip-Franlab. (2016). *FracaFlow user's guide.* Rueil Malmaison, France.
9. Berre, I., Doster, F., & Keilegavlen, E. (2019). Flow in fractured porous media: A review of conceptual models and discretization approaches. *Transport in Porous Media, 130*(1), 215–236.
10. Botros, F. E., Hassan, A. E., Reeves, D. M., & Pohll, G. (2008). On mapping fracture networks onto continuum. *Water Resources Research 44*(8).

11. CMG. (2012). *Advance compositional and GHG reservoir simulator user's guide*. Calgary, Alberta.
12. Casciano, C., Ruvo, L., Volpi, B., & Masserano, F. (2004). Well test simulation through Discrete Fracture Network modelling in a fractured carbonate reservoir. *Petroleum Geoscience, 10*(4), 331–342.
13. Choi, M., Seo, J., Park, H., & Sung, W. (2013). Analysis of oil flow in fractured oil reservoir using carbon dioxide (CO_2) foam injection. *Journal of Petroleum and Gas Engineering, 4*(6), 143–144.
14. Crawshaw, J. P., & Boek, E. S. (2013). Multi-scale imaging and simulation of structure, flow and reactive transport for CO_2 storage and EOR in carbonate reservoirs. *Reviews in Mineralogy and Geochemistry, 77*(1), 431–458.
15. Daniau, F., Aug, C., Lemaux, T., Lalou, R., & Lemaire, O. (2008). An innovative and multi-disciplinary methodology for modelling naturally fractured reservoirs. In *Proceedings of the 70th EAGE Conference and Exhibition Incorporating SPE EUROPEC*, Rome, Italy.
16. De Dios, J. C., Le Gallo, Y., & Marín, J. A. (2019). Innovative CO_2 injection strategies in carbonates and advanced modeling for numerical investigation. *Fluids, 4*(1), 52.
17. De Dios, J. C., Delgado, M. A., Marín, J. A., Salvador, I., Álvarez, I., Martinez, C., & Ramos, A. (2017). Hydraulic characterization of fractured carbonates for CO_2 geological storage: Experiences and lessons learned in Hontomín Technology Development Plant. *International Journal of Greenhouse Gas Control, 58*, 185–200.
18. Ellis, B., Peters, C., Fitts, J., Bromhal, G., McIntyre, D., Warzinski, R., et al. (2011). Deterioration of a fractured carbonate caprock exposed to CO_2-acidified brine flow. *Greenhouse Gases: Science and Technology, 1*(3), 248–260.
19. Ettehadtavakkol, A., Lake, L. W., & Bryant, S. L. (2014). CO_2-EOR and storage design optimization. *International Journal of Greenhouse Gas Control, 25*, 79–92.
20. Gale, J. (2004). Geological storage of CO_2: What do we know, where are the gaps and what more needs to be done? *Energy, 29*(9–10), 1329–1338.
21. Le Gallo, Y., & De Dios, J. C. (2018). Geological model of a storage complex for a CO_2 storage operation in a naturally fractured carbonate formation. *Geosciences, 8*(9), 354–367.
22. Gilman, J. R., & Kazemi, H. (1988). Improved calculations for viscous and gravity displacement in matrix blocks in dual-porosity simulators (includes associated papers 17851, 17921, 18017, 18018, 18939, 19038, 19361 and 20174). *Journal of Petroleum Technology, 40*(01), 60–70.
23. Hardebol, N. J., Maier, C., Nick, H., Geiger, S., Bertotti, G., & Boro, H. (2015). Multiscale fracture network characterization and impact on flow: A case study on the Latemar carbonate platform. *Journal of Geophysical Research: Solid Earth, 120*(12), 8197–8222.
24. Hosseininoosheri, P., Hosseini, S., Nuñez-López, V., & Lake, L. (2018). Impact of field development strategies on CO_2 trapping mechanisms in a CO_2–EOR field: A case study in the permian basin (SACROC unit). *International Journal of Greenhouse Gas Control, 72*, 92–104.
25. Hutcheon, I., Shevalier, M., Durocher, K., Bloch, J., Johnson, G., Nightingale, M., et al. (2016). Interactions of CO_2 with formation waters, oil and minerals and CO_2 storage at the Weyburn IEA EOR site, Saskatchewan, Canada. *International Journal of Greenhouse Gas Control, 53*, 354–370.
26. De Joussineau, G., Barrett, K. R., Alessandroni, M., Le Maux, T., & Leckie, D. (2016). Organization, flow impact and modeling of natural fracture networks in a karstified carbonate bitumen reservoir: An example in the Grosmont Formation of the Athabasca Saleski leases, Alberta, Canada. *Bulletin of Canadian Petroleum Geology, 64*, 291–308.
27. Kazemi, H. (1969). Pressure transient analysis of naturally fractured reservoirs with uniform fracture distribution. *Society of Petroleum Engineers Journal, 9*(04), 451–462.
28. Kazemi, H., Merrill, L., Jr., Porterfield, K., & Zeman, P. (1976). Numerical simulation of water-oil flow in naturally fractured reservoirs. *Society of Petroleum Engineers Journal, 16*(06), 317–326.
29. Lange, A., & Bruyelle, J. (2011). A multimode inversion methodology for the characterization of fractured reservoirs from well test data. In *SPE EUROPEC/EAGE Annual Conference and Exhibition*. SPE-143518, Vienna, Austria.

30. Laochamroonvorapongse, R., Kabir, C., & Lake, L. W. (2014). Performance assessment of miscible and immiscible water-alternating gas floods with simple tools. *Journal of Petroleum Science and Engineering, 122,* 18–30.
31. Lee, L. (1982). Well testing. In *SPE Monograph Series*. Texas: SPE.
32. Li, L., & Voskov, D. (2018). Multi-level discrete fracture model for carbonate reservoirs. In *ECMOR XVI-16th European Conference on the Mathematics of Oil Recovery* (pp. 1–17). European Association of Geoscientists & Engineers.
33. Liu, R., Li, B., Jiang, Y., & Huang, N. (2016). Mathematical expressions for estimating equivalent permeability of rock fracture networks. *Hydrogeology Journal, 24*(7), 1623–1649.
34. March, R., Doster, F., & Geiger, S. (2018). Assessment of CO_2 storage potential in naturally fractured reservoirs with dual-porosity models. *Water Resources Research, 54*(3), 1650–1668.
35. Mishra, S., Kelley, M., Slee, N., Gupta, N., Bhattacharya, I., & Hammond, M. (2013). Maximizing the value of pressure monitoring data from CO_2 sequestration projects. *Energy Procedia, 37,* 4155–4165.
36. Mishra, S., Ravi Ganesh, P., Kelley, M., & Gupta, N. (2017). Analyzing the performance of closed reservoirs following CO_2 injection in CCUS projects. *Energy Procedia, 114,* 3465–3475.
37. Mishra, S., Ravi Ganesh, P., Pasumarti, A., Gupta, N., & Pardini, R. (2018). Practical reservoir engineering metrics for analyzing the performance of CCUS projects. In *Proceedings of 14th International Conference on Greenhouse Gas Control Technologies, GHGT-14*, October 21st–25th 2018, Melbourne, Australia.
38. Moinfar, A. (2013). Development of an efficient embedded discrete fracture model for 3D compositional reservoir simulation in fractured reservoirs.
39. Nakashima, T., Arihara, N., & Sato, K. (2001). Effective permeability estimation for modeling naturally fractured reservoirs. In *SPE Middle East Oil Show*. Society of Petroleum Engineers.
40. Ngo, T. D., Fourno, A., & Noetinger, B. (2017). Modeling of transport processes through large-scale discrete fracture networks using conforming meshes and open-source software. *Journal of Hydrology, 554,* 66–79.
41. Ouenes, A., Anderson, T. C., Klepacki, D., Bachir, A., Boukhelf, D., Robinson, G. C., et al. (2010). Integrated characterization and simulation of the fractured Tensleep reservoir at Teapot Dome for CO_2 injection design. In *SPE Western Regional Meeting*. Society of Petroleum Engineers.
42. Palacio, J. C., & Blasingame, T. A. (1993). Decline curve analysis using type curves—Analysis of gas well production data. SPE paper 25909 presented at the 1993 Joint Rocky Mountain Regional/Low Permeability Reservoirs Symposium, Denver, CO, USA, April 26–28, 1993.
43. Panfili, P., & Cominelli, A. (2014). Simulation of miscible gas injection in a fractured carbonate reservoir using an embedded discrete fracture model. In *Abu Dhabi International Petroleum Exhibition and Conference*. Society of Petroleum Engineers.
44. Patil, V. V., McPherson, B. J., Priewisch, A., Moore, J., & Moodie, N. (2017). Factors affecting self-sealing of geological faults due to CO_2-leakage. *Greenhouse Gases: Science and Technology, 7*(2), 273–294.
45. Peck, W. D., Azzolina, N. A., Ge, J., Bosshart, N. W., Burton-Kelly, M. E., Gorecki, C. D., et al. (2018). Quantifying CO_2 storage efficiency factors in hydrocarbon reservoirs: A detailed look at CO_2 enhanced oil recovery. *International Journal of Greenhouse Gas Control, 69,* 41–51.
46. Peng, D.-Y., & Robinson, D. B. (1976). A new two-constant equation of state. *Industrial & Engineering Chemistry Fundamentals, 15*(1), 59–64.
47. Ray, D. S., Al-Shammeli, A., Verma, N. K., Matar, S., De Groen, V., De Joussineau, G., et al. (2012). Characterizing and modeling natural fracture networks in a tight carbonate reservoir in the Middle East: A methodology. *Bulletin of the Geological Society of Malaysia, 58,* 29–35.
48. Raziperchikolaee, S., Alvarado, V., & Yin, S. (2014). Microscale modeling of fluid flow-geomechanics-seismicity: Relationship between permeability and seismic source response in deformed rock joints. *Journal of Geophysical Research: Solid Earth, 119*(9), 6958–6975.
49. Raziperchikolaee, S., & Mishra, S. (2019). Numerical simulation of CO_2 huff and puff feasibility for light oil reservoirs in the Appalachian Basin: Sensitivity study and history match of a CO_2 pilot test. *Energy & Fuels, 33*(11), 10795–10811.

50. Raziperchikolaee, S., Babarinde, O., Sminchak, J., & Gupta, N. (2019a). Natural fractures within Knox reservoirs in the Appalachian Basin: Characterization and impact on poroelastic response of injection. *Greenhouse Gases: Science and Technology, 9*(6), 1247–1265.
51. Raziperchikolaee, S., Kelley, M., & Gupta, N. (2019b). A screening framework study to evaluate CO_2 storage performance in single and stacked caprock–reservoir systems of the Northern Appalachian Basin. *Greenhouse Gases: Science and Technology, 9*(3), 582–605.
52. Ren, B., & Duncan, I. (2019). Modeling oil saturation evolution in residual oil zones: Implications for CO_2 EOR and sequestration. *Journal of Petroleum Science and Engineering, 177,* 528–539.
53. Rinaldi, A. P., & Rutqvist, J. (2013). Modeling of deep fracture zone opening and transient ground surface uplift at KB-502 CO_2 injection well, In Salah, Algeria. *International Journal of Greenhouse Gas Control, 12,* 155–167.
54. Rinaldi, A. P., Rutqvist, J., Finsterle, S., & Liu, H.-H. (2017). Inverse modeling of ground surface uplift and pressure with iTOUGH-PEST and TOUGH-FLAC: The case of CO_2 injection at In Salah, Algeria. *Computers & Geosciences, 108,* 98–109.
55. Rinaldi, A. P., Rutqvist, J., Finsterle, S., & Liu H. (2014). Forward and inverse modeling of ground surface uplift at In Salah, Algeria. In *48th US Rock Mechanics/Geomechanics Symposium.* American Rock Mechanics Association.
56. Terry, R. E., & Rogers, J. B. (2014). *Applied petroleum reservoir engineering* (3rd ed.). Prentice Hall.
57. Verscheure, M., Fourno, A., & Chilès, J. P. (2012). Joint inversion of fracture model properties for CO_2 storage monitoring or oil recovery history matching. *Oil & Gas Science and Technology-Revue D'IFP Energies Nouvelles, 67,* 221–235.
58. Vilarrasa, V., Rinaldi, A. P., & Rutqvist, J. (2017). Long-term thermal effects on injectivity evolution during CO_2 storage. *International Journal of Greenhouse Gas Control, 64,* 314–322.
59. Warren, J., & Root, P. J. (1963). The behavior of naturally fractured reservoirs. *Society of Petroleum Engineers Journal, 3*(03), 245–255.
60. Wu, Y.-S. (2015). *Multiphase fluid flow in porous and fractured reservoirs.* Gulf Professional Publishing.
61. Zhang, N., Yin, M., Wei, M., & Bai, B. (2019). Identification of CO_2 sequestration opportunities: CO_2 miscible flooding guidelines. *Fuel, 241,* 459–467.

Risk Assessment and Mitigation Tools

A. Hurtado, S. Eguilior, J. Rodrigo-Naharro, L. Ma, and F. Recreo

Abstract Developing engineering projects involving geological systems, such as the Carbon Capture and Storage technologies (CCS), is a complex task with significant challenges. Often the subsoil is poorly investigated and projects often face difficult management of risk components related to uncertainties in the geological environment. Understanding and assessing the environmental risks in these projects should provide satisfactory answers to questions regarding whether CO_2 can leak and what would happen, specifically regarding the consequences for safety, health and the environment. It is worth noting the importance of giving an adequate answer to these questions, among other reasons, due to its influence on the public acceptance of this technology. There is a clear relationship between the early estimation of environmental risks and the social acceptance of technologies. This allows overcome both mistrust and erroneous concepts that citizens could have in relations to them. As indicated in Guide 1 for the application of the European CCS Directive, the environmentally safe management of CO_2 geological storage must be a fundamental objective in any project associated with CCS processes. All this has to be integrated with monitoring strategies for verifying the behavior of the site.

Keywords Risk assessment · Monitoring CO_2 · Bayesian Networks · SRF methodology · Mitigation

1 Introduction

Developing engineering projects involving geological systems, such as the Carbon Capture and Storage (CCS) technologies, is a complex task with significant challenges. Often the subsoil is poorly investigated and projects often face difficult

A. Hurtado · S. Eguilior (✉) · J. Rodrigo-Naharro · F. Recreo
CIEMAT, Avda. Complutense 40, 28040 Madrid, Spain
e-mail: sonsoles.eguilior@ciemat.es

L. Ma
School of Mines and Energy, Technical University of Madrid, Calle de Rios Rosas 21, 28003 Madrid, Spain

J. C. de Dios et al. (eds.), *CO_2 Injection in the Network of Carbonate Fractures*, Petroleum Engineering, https://doi.org/10.1007/978-3-030-62986-1_7

management of risk components related to uncertainties in the geological environment. Understanding and assessing the environmental risks in these projects should provide satisfactory answers to questions regarding whether CO_2 can leak and what would happen, specifically regarding the consequences for safety, health and the environment [1]. It is worth noting the importance of giving an adequate answer to these questions, among other reasons, due to its influence on the public acceptance of this technology. There is a clear relationship between the early estimation of environmental risks and the social acceptance of technologies, since a reasonable guarantee that the society could benefit of the use of these technologies avoiding secondary negative effects is pursued. This allows overcome both mistrust and erroneous concepts that citizens could have in relations to them.

Both safety and long-term risk management of CO_2 Geological Storage (CGS) should be considered as a part of a continuous and iterative process throughout the life cycle of the project, which, based on appropriate methodologies, has to establish a robust and reliable framework that should identify, evaluate and manage both risks and uncertainties in each of the associated phases of the project, including: (i) the identification and early selection of geological formations; (ii) its characterization; (iii) the development of the project; (iv) the operating period; (v) the post-closure operations in the pre-transfer phase of control of the facility; and (vi) the transfer of responsibilities. During all of them, risk management will aim at continuous improvement in the knowledge of the system and its associated risks in order to help attaining project objectives. As indicated in Guide 1 for the application of the European CCS Directive [2] the environmentally safe management of CGS must be a fundamental objective in any project associated with CCS processes, which must be present in all phases of the project.

Within the different focuses and methodologies it will be necessary to reflect, know and take into account the positive aspects of each one of them, as well as its limitations in order to get the best out of each one in the different phases of its development. Thus, for example, already from the first phase, consisting of the site selection, the need to incorporate risk management (RM) arises and it will be an activity that will require specific research [3] for the development of methodologies that allow applying a systemic point of view and tools that enable the integration of available knowledge and the treatment of the high uncertainties associated with these initial phases. All this has to be integrated with monitoring strategies for verifying the behavior of the site.

The main objectives of the monitoring applied to a CGS, are those related to: (i) the control of the storage operation (e.g. capacity, injectivity, containment); (ii) the control of the risks associated with possible CO_2 leakages (e.g. contamination of shallower aquifers, escapes to the surface); and (iii) the calibration of the numerical models simulating the behavior of CO_2 for the long term to estimate the evolution of both risks and operation as accurate as possible. To achieve these objectives, monitoring systems should cover three aspects: (i) monitoring the operation of injection; (ii) monitoring for verification (location, distribution and migration of CO_2, integrity of wells and seal formation); and (iii) monitoring the environment [4].

Finally, monitoring is intimately related to risk analysis and mitigation or remediation measures. In this sense, risk assessments should provide the basis for those measures, which are aimed to prevent any risk to the environment or human health in case of CO_2 leakages from a geological storage of CO_2.

2 General Elements of Risk Analysis and Assessment Methodologies

At the international level, and among the different directives and guides [5], there is a broad consensus on the definition of «risk». A typical definition in the field of engineering project management is one that qualifies Risk as any uncertain event o condition such that, if it occurs, it has an effect—either positive or negative—on a project objective [6]. The exact wording of the different definitions may vary, but they all coincide in the definition from two components. The first one is referred to "uncertainty", since risk is something not materialized which may or may not occur. The second one refers to what would happen if said risk were to materialize, that is, its impact or consequences, since risks are always defined in terms of their effect on the objectives of the project.

Risk Management (RM) tools allow to face the knowledge and control of the same in a wide variety of human activities, industrial or not. Thus, the RM allows structuring the effort of an organization to identify, measure, classify and assume, eliminate, mitigate, transfer, or control the different levels of risks associated with a project. Figure 1 shows a possible structure of a RM process. The different phases are general for all management systems, although their framing may vary among methodologies, and should be considered as part of a continuous and iterative process throughout the life cycle of the project. A fundamental aspect is the need to ensure the identification of all significant risks, from which the corresponding measures can be taken (risk analysis). An unidentified risk allows neither its evaluation nor its monitoring, reduction or cancellation. After the analysis phase, the risk evaluation phase can be considered through which the severity of consequences of risk materialization—it previously identified in the risk analysis phase—and the probabilities associated with said materialization could be estimated and, based on adequate methodologies, to establish a robust and reliable framework that allows to evaluate both consequences and uncertainties in each of the phases of the project.

2.1 *Analysis and Evaluation of Environmental Risks in Geological Storage of CO_2*

The risk analysis and subsequent risk assessment process should be tailored to the relevant stage of development of the project, reflecting the decisions to be made and

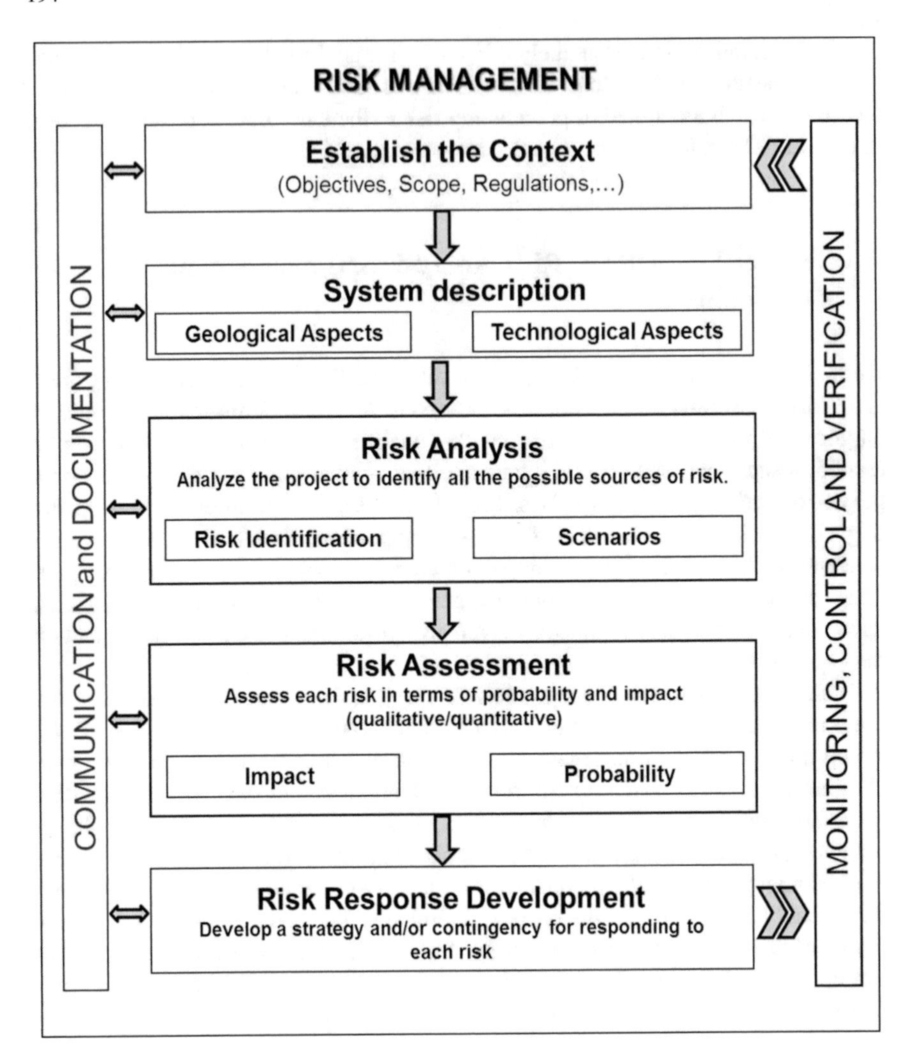

Fig. 1 Risk management steps

the level of available detailed information. In addition, it must be noted that no one project is the same as another [7] due to variations imposed by the geology of each site and its behavior in connection with the process of CO_2 injection. Thus, the level of risk will vary from one site to another, i.e., it is not advisable to take decisions based on an a priori general risk prioritization.

2.1.1 Risk Analysis

The objective of this phase is the identification of all the risks that may directly or indirectly give rise to undesired consequences in the project. In this context, consideration will have to be given to:

- The features of the different elements that make up the system.
- Events and processes both internal and external to the system.

The risks identified will depend on the context of the evaluation: objectives, premises, scope, regulations, spatial and/or temporal limits, and so on.

At this stage, the elements (characteristics with their properties, chemical and physical processes, events than can alter its normal evolution) which affect behavior and evolution of the system are identified and classified.

The main issue at this phase is whether it is possible to ensure that the set of risks is complete. It is impossible to demonstrate strictly but a review process open to broad groups in the scientific community is probably the best way to reasonably ensure that the risk analysis is complete. So, it is important:

- To follow an approach that allows us to guarantee and defend that the list obtained is sufficient for the evaluation that is being carried out.
- To document all judgments and their reasoning.
- To be iterative and flexible.
- To allow a systematic and orderly visualization of the system.

In the risk analysis phase different sources are usually used (e.g. literature review, expert elicitation, historical records or experience gained in analogous disciplines). In addition, different systematic approaches [8] are available, such as the FEP (Features, Events and Processes) methodology that identifies the characteristics, events, and project specific processes that are used to explore the sources of project risks and to generate a comprehensive range of evolution scenarios thereof; failure trees, used to identify risk scenarios; or the Failure Mode and Effect Analysis (FMEA) methodology for failure mode analysis and its effects [9, 10].

An important aspect in assessing long-term risks of a project is the identification of the possible scenarios of evolution of the system. The need to conduct a scenario development in performance and risk assessments arises from the fact that it is virtually impossible to accurately predict the evolution of the system over time.

The scenarios development phase aims to achieve a set of illustrative scenarios of system behavior through time to provide a reasonably complete picture of the evolutionary paths of the system. These scenarios shall define the context, in broad terms, in which to perform the steps of modelling and consequence analysis since, in order to quantify the potential impacts and risks associated with the project, one needs to assess its possible long-term behavior in the geological medium as well as to define possible migration pathways and mechanisms, that will depend on the scenario under consideration.

Among systematic methodologies used to develop scenarios, one can mention the systems analysis approach, which includes FEPs analysis methodology, successfully applied in the field of radioactive waste disposal to assess the problem of long-term radioactive waste behavior in geological media [11] and which is also the approach adopted, for example, within the CGS performance and security evaluation in the Weyburn project [12].

2.1.2 Risk Assessment

Once the risks have been identified, it will be necessary to assign values to each of the identified failure scenarios (probability) and to the impacts on each initially defined objective (impact function). The total risk of the system will be the sum of the probability of each scenario by its impact function (Eq. 1).

$$Risk = \sum_{i=1}^{N} (Probability)_i \cdot (Impact)_i; \quad N\, is\, the\, number\, of\, scenarios \qquad (1)$$

For risk assessments to be consistent and meaningful, the application of appropriate methodologies in the evaluation of probability and impact is essential. Assessment methodologies can be divided into two broad categories: qualitative and quantitative. Technological maturity or gaps in knowledge in the evolution of disturbed natural systems, as well as the project phase, determine the nature of the assessment to be used.

In the qualitative ones, the assignation of probabilities and impacts is made through significance levels. When there is a lack of specific information and/or knowledge, a qualitative risk assessment can be sufficiently effective. Qualitative approaches classify risks through scores that allow them to be compared. They often use qualitative methods to assign estimates of probability and/or consequences, and then use quantitative tools to classify and evaluate them in more detail. They can serve as a platform towards a quantitative system, particularly when detailed data is lacking, and can be used as a means to capture subjective opinions, open discussions, and become in a framework for identifying where an additional analytical effort is required.

The most common qualitative methods are: the two-dimensional Probability—Impact matrix, the Bow-Tie diagrams [13], the Vulnerability Evaluation Framework (VEF), the Structured "What-If" Techniques (SWIFT), the Multi-Criteria Assessment (MCA) [8, 14], and the Selection and Classification Framework or Screening and Ranking Framework (SRF) [15]. This latter one has been satisfactorily used in early environmental risks assessments focused on its effects on Health, Safety and the Environment (HSE) [16], for the selection of possible CO_2 geological storage sites [17]—a clear example of a geo-project with an important limitation both in initial data and in knowledge about the evolution and consequences of disturbed natural systems. Among qualitative methodologies, Expert Judgment (EJ) constitutes an essential tool used to request informed judgments based on the training and experience of experts.

The quantitative risk assessment develops numerical estimates of the probability of occurrence and of the magnitude of the impact in the different scenarios. The quantitative approaches have used approaches for uncertainty treatment based on EJ combined with risk matrices (e.g. Schlumberger's Carbon Workflow), evidence supported logic (e.g. CO2TESLA) and Bayesian Networks (BN) [18]. In quantitative approaches, these methodologies are combined with specific software codes for calculating impacts, and they are applied through performance assessment models which, based on a global view of the system, provide the ability to simulate the dynamic evolution of the entire system (e.g. CO2-PENS, Certification Framework, QPAC-CO2 [15], ABACO2G (Aplicación de Bayes al Almacenamiento de CO_2 Geológico) [19] or NRAP-IAM [20]) or parts of it, such as wells, or impacts on aquifers in case of a leakage [9].

Quantitative methods are used in well-known systems, where the level of uncertainty is relatively low, and use approaches that directly address uncertainties. They measure the credibility of a hypothesis based on the evidence that supports it. They can be represented by a probability density function, if the frequentist concept of probability is used, or make use of the uncertain or approximated reasoning, related with fuzzy logic or similar models. The approaches used by the latter may be grouped as follows: Empirical (MYCIN, Prospector); Approximated Methods; Diffuse Logic; Dempster-Shafer Theory and BN.

3 Risk Assessment of a CO_2 Geological System

This section presents the risk analysis and evaluation process in the initial selection and characterization stages of a site, from the perspective of formal risk analysis. It is designed with the aim of developing the methodologies and technologies that facilitate the CGS in low permeability and fractured carbonate formations (limestones, dolomites, and carnioles of the Lower Jurassic), the primary objective for the development of CCS technologies in Spain, as these lithologies possibly have the greatest potential for geological storage in Spain.

Once the criteria and performance indicators had been defined [21], the first step was to carry out a risk analysis and evaluation of the possible locations where a CO_2 storage system could be located. This allowed to classify the zones from the point of view of their environmental risks and to help in the selection of a site [17]. Once the site was selected as an initial step for the risk evaluation, the main leakage scenarios [22–24] were identified, namely:

- Leakage through wells.
- Leakage due to fracturing of the seal rock due to overpressurization.
- Leakage through the seal rock pore system, either due to overpressures or the presence of an undetected area of high permeability.
- Leakage through an existing fault.
- Migration of the brine from the formation.

Later on, a methodology was developed and applied to evaluate the risks associated with them [18]. It is a probabilistic approach that allows us to explicitly deal with the uncertainties associated with the ranges of variability of the parameters, the scenarios and the conceptual models of the processes involved in each scenario. To do this, an integrated tool was developed and implemented that has allowed addressing the fate and effects of the injected CO_2, also including uncertainties in the predictions.

3.1 Application of the Environmental Evaluation of HSE Risks in the Site Selection Phase

Selecting a safe site, capable of sequestering CO_2 for long periods of time and with minimal risk is the first step in a Geological CO_2 Storage project, and it requires specific research [3]. In this case the methodology developed by Oldenburg [25] has been applied to three candidate areas for the location of a pilot CO_2 injection plant in the western part of the Basque-Cantabrian Basin: Huérmeces, Huidobro and Leva, in the Burgos province of Spain (Fig. 2).

The methodology makes use of the available information of qualitative type (studies, reports, publications, EJ) as an approximation for the evaluation of possible combinations of probabilities and consequences. Many of the properties and values considered in these early phases involve estimates that can be measured and modeled in later phases. Given the usual absence of direct data in the early stages of the project, maintaining uncertainty as an input and output value in the methodology is a key condition.

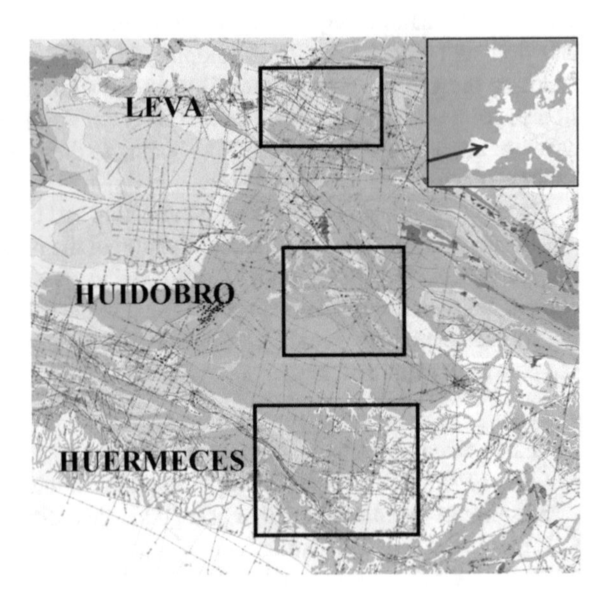

Fig. 2 Study areas

The methodology supports the evaluation of different sites and different scenarios (e.g. related to well technology options, water management, etc.) in one or more specific locations. This process allows us to compare different options, which in turn facilitates the decision-making process. Furthermore, this approach constitutes a powerful communication tool to inform stakeholders through knowledge sharing and, in particular, about the assessed risks.

The methodology is flexible and can be adapted to the different types of projects where globally it will allow evaluating the main aspects related to their safety, including those focused on: (1) the natural properties of the site and (2) the technological properties of the project. The main aspects related to risk are described according to their characteristics (c_i), that is, the fundamental parts into which the project can be divided from the point of view of its HSE risks. These, in turn, are broken down into attributes (a_i), which determine how characteristic ci is competent in fulfilling its HSE risk-oriented function. Finally, these attributes are divided into properties (p_i), based on whose values the performance of the attributes with respect to HSE risks will be determined.

Table 1 shows the characteristics and attributes associated with an assessment of the risks of a CO_2 Storage and Table 2 shows an example of the characteristic/attribute/property set.

Properties values entered by the evaluator represent "proxies" or reasonable substitutes for site or technology-related characterization data or modeling results, which may not be known at the time of evaluation. Thus, for example, the "lithology" property of the "Primary Seal" attribute (see Table 2) is used as an indicator of permeability and porosity. The subjacent idea is that permeability and porosity distributions may not be available in the early stages of the project, but lithology gives an initial adequate representation of these properties. Associates uncertainties are entered through confidence values associated with each property. Therefore, each property has two values assigned: one will measure its performance with respect to risk; the other, the evaluator's confidence in the assigned value. These allocations,

Table 1 Characteristics and attributes for a geological CO_2 storage system. The HSE risk of the system will be evaluated based on the values and uncertainties associated with each of them

Geological storage of CO_2	
Characteristics	Attributes
Potential for primary containment	Primary seal
	Depth
	Reservoir
Potential for secondary containment	Secondary seal
	Shallower seals
Attenuation potential	Surface characteristics
	Groundwater hydrology
	Existing wells
	Faults

Table 2 Example of a group of characteristics/attributes/properties, as well as the risk element to which it is associated in a CO_2 geological storage [25]

Characteristics	Attributes	Properties	Proxy for
Potential for primary containment	Primary seal	Thickness	Likely sealing effectiveness
		Lithology	Permeability, porosity
		Demonstrated sealing	Leakage potential
		Lateral continuity	Integrity and spill point
	Depth	Distance below surface	Density of CO_2 in reservoir
	Reservoir	Lithology	Likely storage effectiveness
		Permeability and porosity	Injectivity, capacity
		Thickness	Areal extent of injected plume
		Fracture or primary porosity	Migration potential
		Pore fluid	Injectivity, displacement
		Pressure	Capacity, tendency to fracture
		Tectonics	Induced fracturing, seismicity
		Hydrology	Transport by groundwater
		Deep wells	Likelihood of well pathways
		Fault permeability	Likelihood of fault pathways

together with the available information and the adopted decisions, should be included in the evaluation to allow transparency and traceability of the process [25].

The methodology makes use of the "multiple barrier system" concept, widely developed in research on ensuring the safety of systems involving geological media, such as the geological storage of radioactive wastes [26]. Thus, in anticipation of a failure in the primary containment system, it is necessary to evaluate the attenuation capacity of HSE impacts by the secondary levels of the geological system, and the possibilities of attenuation of impacts must also be examined and evaluated, for instance, the fast dispersion in the atmosphere of possible contaminants or their mixture with geological/natural/environmental waters up to safe levels, as well as the reaction times to reach dangerous concentrations [27]. All this will depend on the characteristics of both the contaminants and site location and land surface.

The main benefit of the methodology is that it formally expresses both the knowledge and the associated uncertainties, so that in future iterations it could be revisited and modified should new data were available.

The system supports a wide degree of versatility, allowing the evaluator to assign different weights based on the relative importance for risk of the different characteristics/attributes/properties. The transparency of the system and its simplicity allows any reviewer to modify the assigned weights and perform further analyses to compare the effects of those changes on the system response. The results of the methodology allow, on the one hand, to compare the risks associated with different locations (or different scenarios for the same site), as it can be seen in Fig. 3. In addition, it is also possible to examine the relationships between the evaluations of the attributes and their certainties, establishing comfort zones and zones where the attributes should improve their characterization (see Fig. 3). Finally, it should be noted that the safety areas of system operation will be defined, in much more advanced phases, by the values of the system's fundamental behavior indicators (or Key Performance Indicators—KPI [28]), associated with monitoring activities, which is not feasible in the earliest stages.

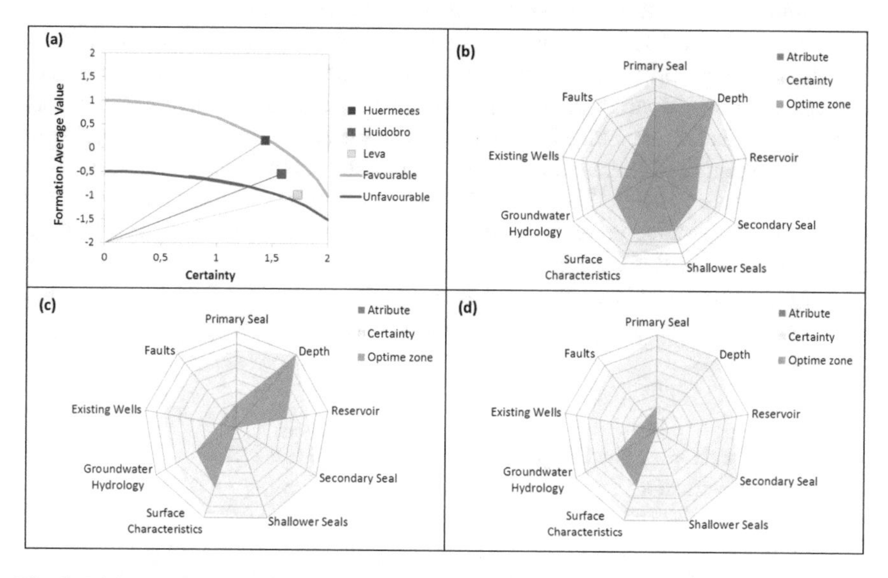

Fig. 3 Risks associated with different alternative sites for a CO_2 geological storage project (**a**). Valuations of different attributes and their uncertainties for the Huermeces (**b**), Huidodro (**c**) and Leva (**d**) sites

3.2 *Environmental Risks Assessment Using Bayesian Networks*

This section documents the activities which have been carried out in order to move forward to a quantitative estimation risk model. The advanced model is based on the determination of the probabilistic risk component of a geological storage of CO_2 using the formalism of BN. To this end, the first step was to define a BN for the evaluation of system's risks. The behavior of the network was validated with qualitative calculations through comparisons with the results of the SRF methodology. Subsequently, quantitative models were included: the time evolution of the CO_2 plume during the injection period, the time evolution of the drying front, the evolution of the pressure front, decoupled from the CO_2 plume progress front; and the implementation of escape submodels, and leakage probability functions, through major leakage risk elements (fractures/faults and wells/deep boreholes) which together define the space of events to estimate the risks associated with the CGS system. Then a quantitative probability risk functions of the total system CO_2 storage and of each one of their subsystems (storage subsystem and the primary seal; secondary containment subsystem and dispersion subsystem or tertiary one) were obtained.

Bayesian Networks [29] are acyclic directed graphs in which the nodes represent random variables and the arcs represent direct probabilistic dependences between them. They allow the structure of a geological storage system to be represented as a graph of the qualitative interactions that exist in the set of variables to be modeled to estimate the risk of leakage in the storage complex and in each of its subsystems, structures and components. The ad hoc directed graph structure that reflects the causal structure of the storage complex model offers a modular view of the relationships and the interactions that exist between its different variables, which enables to make predictions about the effects due to causes external to the system. Injection scenarios, among others, are the most immediate external causes to a storage system. The BN can be seen in Fig. 4.

3.2.1 Application of the Proposed Methodology to the Zone of Huérmeces

The initial BN model is oriented towards the estimation of the probability of risk of leakage in a CGS from EJ, and therefore from qualitative-type data. This model evaluates the combination of the probability of leakage from the primary containment and from the secondary one, as well as the edaphic capacity of attenuation of those potential escapes. The model takes into account and establishes relationships among the variables that define the storage system.

The application of the proposed methodology was implemented in the Huérmeces zone of the BN model built for estimating the risk of leakage (see in Fig. 5 the BN probability of leakage model at the CGS). The calculated probability range is defined

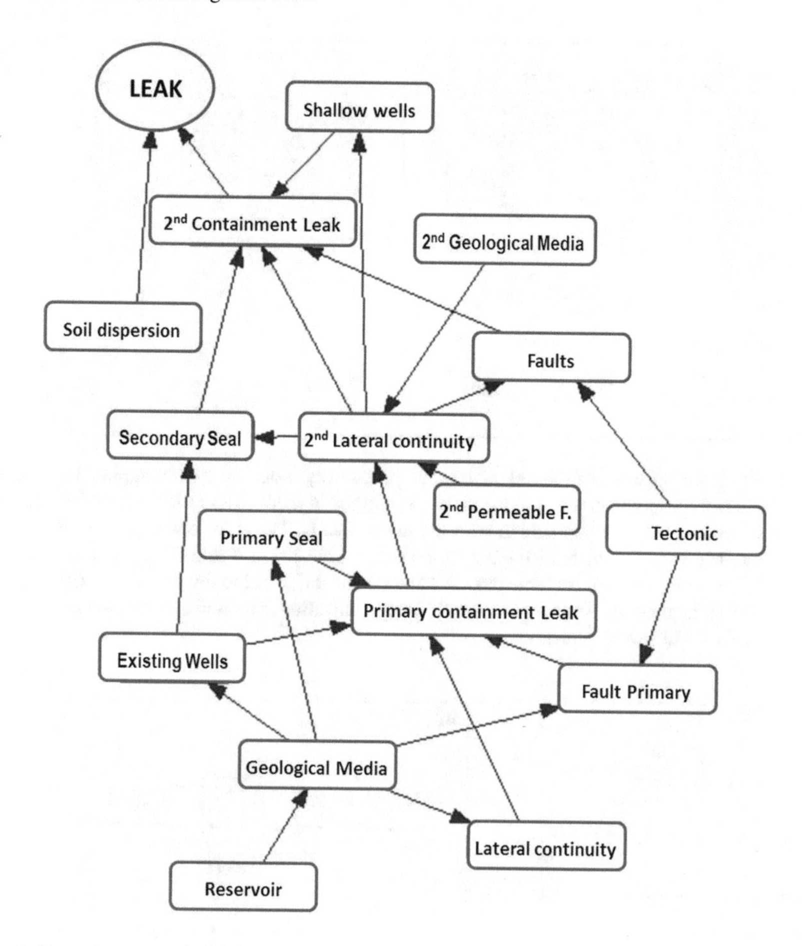

Fig. 4 Bayesian network. Main nodes

by the BN represented in Fig. 5 for the greater (and lower) values of the variability range.

Red and green variables contribute with information to the model. In the BN which determines the upper range (see Fig. 5a), a value of ≈79% is reached, of which, ≈66% indicate a probability trend in favor of leakage. On its turn (see Fig. 5b) obviously the percentage of nodes with information is maintained, but only ≈34% of them indicate probability trends in favor of leakage.

The risk of leakage probability range estimated for the study area is $p \in [0.656, 0.329]$ with a d value of $d = 0.654$. By eliminating from the model those variables related to the edaphic capacity of attenuation of the potential escapes (variables which, at the current stage of development of the project do not give information), the associated probability range is $p \in [0.562, 0.408]$, with a d value of $d = 0.308$ (eliminating variables without information, the uncertainty associated with the calculus diminishes). The results obtained are shown in Fig. 6. The comparison of these

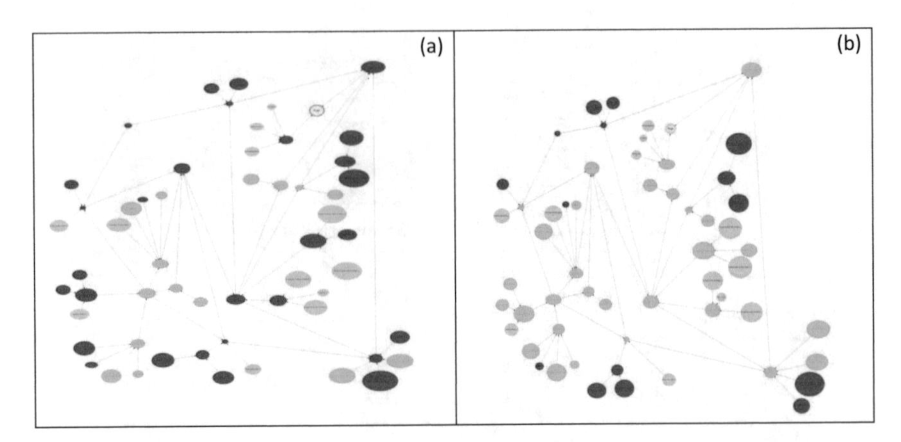

Fig. 5 Bayesian Network of the risk of leakage probability model in a CGS applied to the study area of Huérmeces; **a** corresponds to the network with the higher values of the variability ranges of the variables, and **b** to the application with the lower values. The color code applied refers to the risk probability value estimated for each variable as follows: Red: Probability value greater than 0.5 (its behavior relative to risk is negative), Green: Probability value lower than 0.5 (its behavior relative to risk is positive, a value in favor of safety), and Blue: Probability value equal to 0.5 (its behavior relative to risk is neutral)

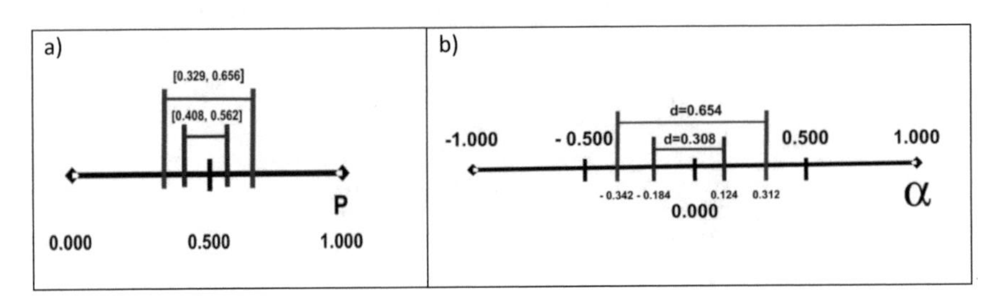

Fig. 6 Graphical representation of results from the application of BN models, the full one and that one without the edaphic capacity, to estimate the probability of the risk of leakage in a CGS applied to the study area of Huérmeces: **a** Probability ranges; **b** "α" and "d" values

results with those obtained in the former evaluation of this same zone with the SCF methodology, seems coherent as both methodologies conclude with a classification of the study zone at an intermediate level of goodness for the CGS, with similar final results in relation to the uncertainties estimation. The BN also allows us to carry out a sensitivity analysis, the results of which can be seen in Fig. 7.

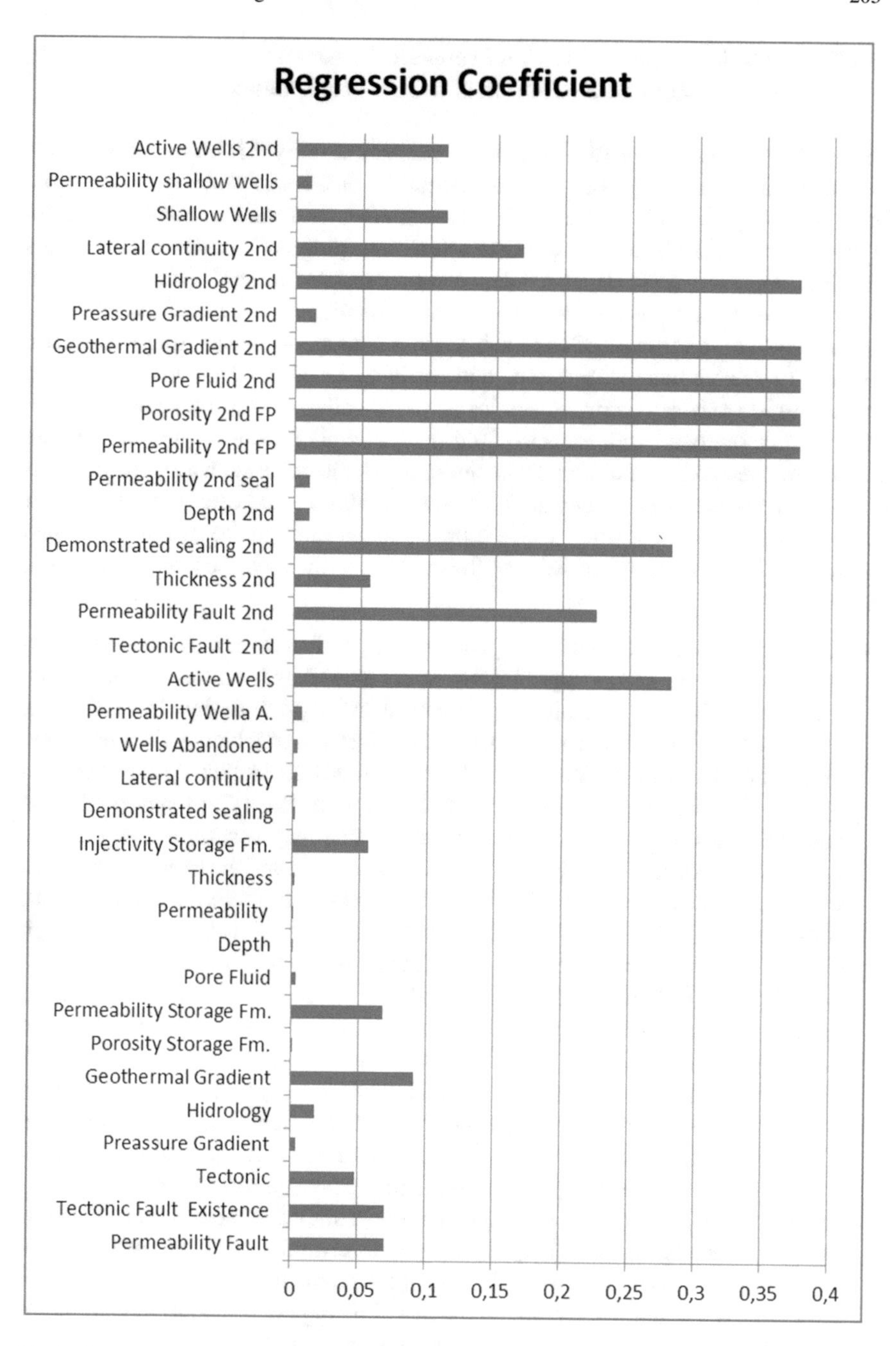

Fig. 7 Results of the sensitivity analysis of the Bayesian Network

3.2.2 Evolution of the BN Model. Probabilistic Model for the Integrated Evaluation of a CGS Performance

In a BN, the estimation of the a priori probabilities by EJ would be only the initial starting point. The Bayesian methodology enables to move gradually from qualitative-type models to quantitative-type models and the combination of both types for probability estimates. Using this flexibility, progress has been made towards the development of a quantitative risk assessment model for the CO_2 injection phase which has enabled to obtain the quantitative probability functions for the total CO_2 storage system, and those of each subsystems (storage—primary seal subsystem; secondary containment subsystem, and tertiary- or dispersive-subsystem). The models used are based on recent studies on the injection of CO_2 into a deep permeable aquifer saturated with brackish fluid from a single injection well, the pressure field generated and the possible leaks through risk elements such as wells or faults [19]. These models are analytical and/or semi-analytical and can be used as a first approximation for calculating leakage probabilities through the above mentioned scenarios. The main characteristics of the models for the scenarios under study will be the following:

- CO_2 plume evolution model: The general scheme of study of a CGS safety corresponds to that of a secular equilibrium system altered by the introduction of CO_2. The injected CO_2 will remain, in its practical entirety and for hundreds of years, as a separate phase enriched with CO_2, the migration of which will be governed by the biphasic flow [CO_2-connate brackish fluid] controlled by injection and hydraulic pressures, and buoyancy associated with density differences. This is due to the fact that the geochemical reactions that may occur between the CO_2 injected, the storage formation rock, the seal formation and the cations in solution in the formation water will take place on time scales of thousands of years [30] since the dissolution of the CO_2 it will be limited by the diffusion and although there may be momentary increases due to local density instabilities, the time scale is on the order of hundreds to thousands of years [31]. From both observations, it appears that during the injection, the displacement is due to a drainage process in which the non-wetting fluid (CO_2) displaces the connate brackish fluid. This shift leaves the connate brackish fluid to residual saturation in the biphasic zone. Hence, the maximum risk of leakage will correspond to the time when CO_2 remains as a separate mobile phase and when the pressures exerted on the medium are high, that is, during the injection period. This is the critical period for risk assessment. For this reason, the first stage of implementing quantitative models for risk assessment is aimed at this phase. The modeling of the evolution of the plume will allow estimating the maximum expected range of the plume for the conditions imposed during the modeling, which is essential in estimating the risk, since it determines the space of events within which are the elements of risk.
- Pressure field model. The injection of CO_2 requires the application of a pressure higher than the storage formation fluid. During the injection operation, the pressures in the aquifer will be distributed radially, from a maximum value located

at the injection well that will decrease almost proportionally to the distance. The necessary overpressure and its area of influence will depend on the receiving aquifer characteristics, its fluids, the amount of CO_2 injected, and the time required for the injection. Applying excessive pressures can lead to hydraulic fracturing of the permeable formation, therefore, for a safe injection operation the maximum admissible injection pressure has to be known. The permanent control of this variable is essential and it will be necessary to anticipate the pressure to which hydraulic fracturing will develop (or movements in fractures) from an estimate of the state of stress to which the formation is subjected at the injection point depth. In sedimentary series the maximum pressure in the vertical direction increases with depth due to the increasing load caused by the increasing thicknesses of rock and fluid. The average value of this increase (lithostatic gradient) is 1 psi/ft (1 lb per square in./foot = 22,620.59 kg m^{-2} s^{-2} $\approx$ 22 MPa km^{-1}), and varies between 22 and 26 MPa km^{-1}. The average hydraulic gradient is 10 MPa km^{-1} or 0.43 psi/ft [32].

- Model of leakage through risk structures. Wells and Boreholes: One of the potential scenarios of risk in a geological storage of CO_2 is the deep wells and boreholes existing in the area of influence of the site, since they can directly put into contact the storage formation with the atmosphere or with shallower aquifers. In this context, it is necessary to differentiate between the CO_2 injection wells, on which a specific regulation that is being developed in various countries would be applied, and other wells already present in the area affected by the CO_2 storage, with characteristics that will depend on its function, year of construction, type of abandonment, etc. Assessing the risk associated with wells will require reliable estimates of both the amount of CO_2 that can migrate through the wells and their probability. In addition, the associated risk will depend on the consequences of said migration, since "risk" implies that the leaked CO_2 may affect a target to be protected and cause harm either to people (in their health, or economic damage), to the environment or to the infrastructure or other assets. In our case, the risk associated with the wells will be determined and integrated into the methodological approach to solve the risk evaluation problems derived from CGS activities, based on the use of BN and Monte Carlo probability. Within this methodology, the "well" model will incorporate the calculation of both the escape rates, which depend fundamentally on the leakage mechanisms, and their probabilities, taking into account all the uncertainties associated with both aspects through Monte Carlo modeling.
- Model of leakage through risk structures. Faults and Fractures: The safe CGS requires that the seal formations can guarantee its long term integrity, this is, the time in which the CO_2 will remain in a supercritical state before entering the dissolved phase as CO_{2aq}. Certain geological structures, especially the faults and fractures that intersect the seal formation and the areas affected by them can suppose preferential paths for the leakage of CO_2. For the purposes of consideration in risk assessment, faults can be considered as two-dimensional conduits whose permeability varies spatially along the fault plane. The permeability in the

direction of the fault is likely to be low in the sections with fill or seal material and the sections in which the fault is clogged will control the flow of CO_2 along it; hence, the importance of taking into account variations in permeability in evaluating the risk associated with faults in a geological storage of CO_2 [33]. In its migration to lower pressure areas, the injected CO_2 may find other fractures connected with the main one, or other permeable formations in which to disperse. In both cases, there will be an attenuation of the ascending flux of CO_2 that must be quantified to estimate the risk. This attenuation will reduce the flow in the fracture, but will also extend the presence of CO_2 in a larger area. Therefore, for a fracture to be considered a risk structure because it constitutes a preferred way of leak it is not necessary for it to reach the surface, it is enough that it may constitute a leakage way to permeable formations of interest such as drinking water aquifers, now or in the future.

Figure 8 shows an example of obtained results from the application of said calculation module where it is possible to visualize stochastic solutions of the dynamic evolutions of the leakage rate by deep well/borehole and by fault/fracture. These results can substitute the probabilities obtained by EJ in the BN and to advance towards quantitative results. In addition, a BN allows us to realize sensitivity analyses and obtaining which parameters introduce more uncertainty in the final results (see Fig. 7). This aspect is essential so that the advance in the characterization of the system is maximizing the benefits in the final reduction of uncertainty.

4 Estimators of the Behavior of a CO_2 Storage Complex

This section is focused on which indicators of the performance of the storage complex and what environmental criteria and security should satisfy the assessment of the long-term risks of a geological storage of CO_2.

The IPCC 2005 [34] classifies the impacts on safety and the environment related to the escape of CO_2 from a geological storage, in two large sections or categories: environmental impacts and on safety of local character, and global effects that could result from the escape of the stored CO_2 into the atmosphere. A CO_2 storage complex should meet the following criteria of acceptability related to CO_2 Containment (global effects) [35] and HSE risk (local effects):

Containment

1. As a design objective for the containment, it is proposed that the mass of CO_2 retained in the storage complex after 1000 years after the injection period is at least 99% of the total CO_2 injected, that is to say that the maximum allowable leakage of mass of CO_2 in 1000 years is less than 1% of the total CO_2 injected.
2. The annual leakage rate corresponding to this containment level is 0.001%/year, which means a retention period of 100,000 years.

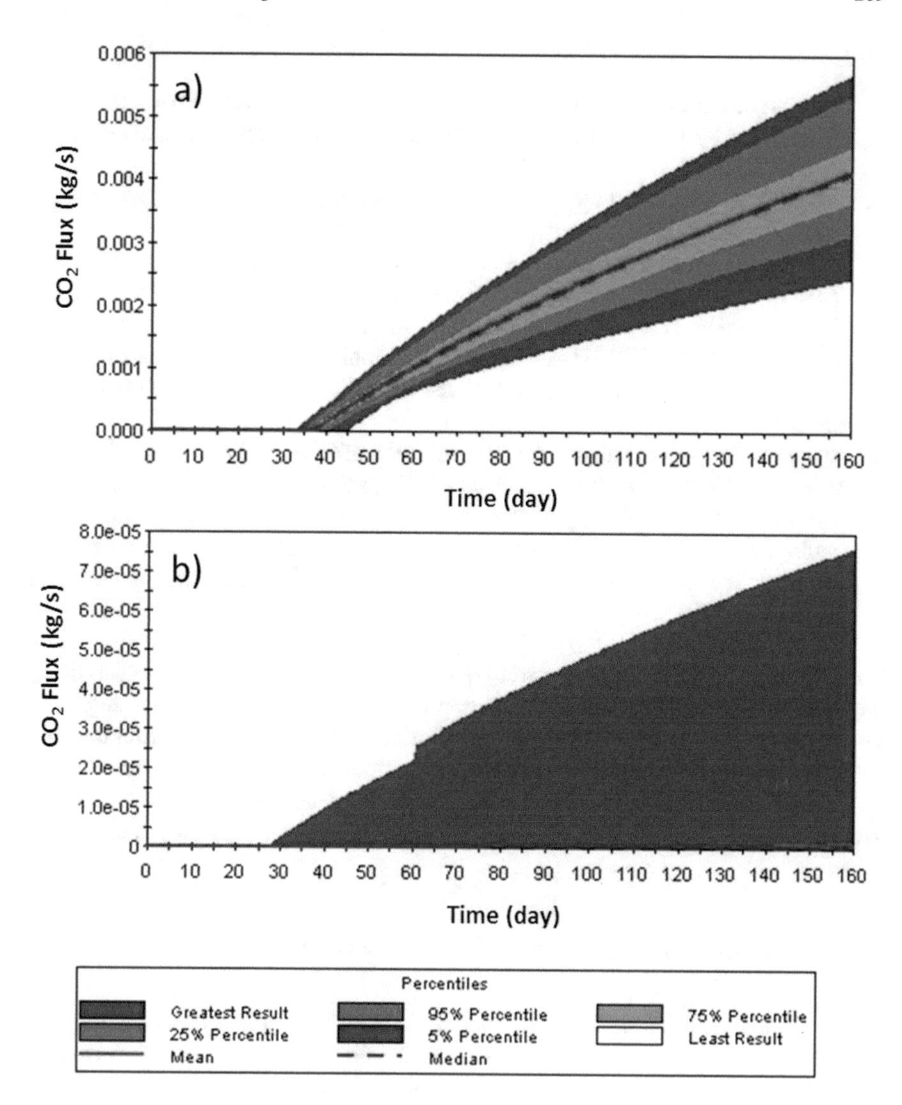

Fig. 8 Probability density functions of CO_2 leakage rates through risk elements: **a** faults; **b** open well

3. The containment is considered to be acceptable if there is a probability greater than or equal to 80% that 99% of the injected mass remains confined in the storage complex during the first 1000 years. That is, a threshold value of leakage risk of 20% of losing the maximum acceptable mass of CO_2.

HSE risk

1. It would be considered unacceptable for the social risk quotient to exceed $1 \cdot 10^{-3}$ deaths/year, by extrapolation of the acceptability criterion for large dams admitted by ANCOLD [36].
2. It would be considered marginally acceptable if that risk quotient was between that value and $1 \cdot 10^{-4}$ deaths/year.
3. Gas-phase CO_2 concentrations in air above the storage complex may not exceed 0.5% or 5000 ppm (Continuous Public Permissible Exposure Limit for 8 h, US Occupational Safety and Health Administration, 1986) or 9150 mg/m^3 according to the Spanish INSHT [37] as a result of the simultaneous action of leakage from storage and the natural emanation of the site determined as a baseline prior to injection.
4. In the shallow soil or subsoil, the concentration of CO_2 must not exceed 5% in air volume and in no case, 20% as a result of CO_2 leakage from the storage complex.
5. The concentration of CO_2 in the dissolved phase in groundwater due to the geological storage of CO_2 should not have an impact on acidification that promotes the mobilization of heavy metals in the aquifers of the underground basin of the site.

Performance criteria are proposed to quantify such functionalities or capabilities, essential for the containment and isolation of CO_2, and to establish the degree of compliance with which the storage complex (storage formation—seal formation) and its subsystems and components must respond to meet the operational requirements. In practice and for regulatory purposes, the performance of a specific geological storage of CO_2 is qualified by **indicators** that assess its degree of acceptability.

4.1 Indicators of Seal Formation Performance

The physical properties of the seal formation that contribute to the isolation of injected fluids under the operating conditions required for storage and on which performance indicators for the storage system can be established are:

Extension and lateral continuity of the seal formation

- It is a necessary condition that the seal geological formation has sufficient extension and lateral continuity to fully cover the area affected by the injection of CO_2 at the moment of the dissipation of the pressure gradients. This area includes both the area occupied by the CO_2 injection and that of the potentially larger area affected by the pressure changes associated with the injection.
- The quantitative criterion of performance to be satisfied is that said area is at least equal to the surface of the maximum extension achievable by the injected CO_2 plume until the dissipation of thermo-hydro-mechanical-chemical (THMQ) gradients.

- In the case of discontinuities, lateral changes of facies, wedges or others, the performance criterion could be satisfied if the discontinuities in a lithological level were in turn sealed by overlapping levels of the confining system of equivalent petrophysical characteristics (porosity and permeability).

Inlet capillary pressure

- The quantitative criterion of performance to be satisfied is that the pressure of the CO_2 column plus the injection pressure is lower than the capillary pressure of the confining system.
- The injection pressure may not exceed a value that may lead to the propagation of fractures in the confining seal formation [38].

Permeability

- The quantitative criterion of performance to be satisfied is that the permeability is less than or equal to that corresponding to lithologies of pelitic fraction (clays, shales or siltstone) measured in mD (milli Darcy), that is, lower than 0.1 mD ($\approx 10^{-12}$ cm^2 of intrinsic permeability, k).

4.2 *Indicators of the Storage Formation Performance*

The physical properties of the storage formation that contribute to the isolation of the injected fluids under the operating conditions required for storage and on which performance indicators for the storage system can be established are:

Thickness and surface area of the injection area

- Although a minimum thickness cannot be established, given that the injection ratio is directly proportional to the average permeability and thickness, a utility value can be given with respect to the thickness that is defined by $permeability \times thickness \geq 10^{-13}$ m^3 [39].

Porosity

- It is the fundamental factor for the storage capacity of the reservoir. The porosity values are usually in a range between 10 and 30% [40]. An optimal storage rock is one with a total porosity value of more than 20%, but total porosities of 12% are perfectly adequate to contain high amounts of CO_2.

Permeability

- Injectivity is directly proportional to the permeability and inversely to the viscosity of the fluid phase injected. One of the interesting characteristics of CO_2 is its low viscosity with respect to that of water (around an order of magnitude). Due to this, the permeability of the formation does not represent a limiting criterion with respect to the injection of CO_2, since the volumetric injection ratios can be important both in formations with high or low permeability values [41].
- A storage formation with a permeability of "good aquifer" (more than 1 m/day), is not necessarily a good storage rock of CO_2 since it makes the control of injected CO_2 difficult. A good storage rock should have effective permeabilities greater than 10 mD and are optimal permeability values of 300 mD.

Injection pressure

- The injection pressures must be greater than 83 bar [42] and will be limited by the tensional state of the seal formation and the reopening of fractures that affect the storage formation. The reservoirs of hydrocarbons with effective traps have a pore pressure gradient value of less than 17.4 kPa/m, which could be considered initially as a safety criterion for the site.
- The sustained pressure will be lower than the fracture pressure, i.e. the pressure at which fractures can be initiated or propagated in the injection zone [38].
- During the injection the pressure in the injection zone cannot exceed 90% of the fracture pressure of the injection zone [38].

5 Monitoring, Verification and Mitigation Tools

Projects for the geological storage of CO_2 should include technical guides for monitoring, verification and accounting of CO_2 stored in geological formations in order to help ensure safe, effective and permanent CGS in the appropriate reservoirs [43]. Monitoring techniques can be applied in atmosphere, near-surface and subsurface to ensure that injected CO_2 remains in the storage formation and that both CO_2 injection process and pre-existing wells do not jeopardized the CO_2 storage complex.

The most usual atmospheric monitoring techniques are optical CO_2 sensors, atmospheric tracers, and eddy covariance flux measurements. On the other hand, near-surface monitoring methods are used to detect potential CO_2 leakages from a CO_2 storage complex, including geochemical monitoring both in the soil and vadose zone and in the near-surface groundwater, surface displacement monitoring, and ecosystem stress monitoring. Furthermore, subsurface monitoring of a CGS project covers a wide range of techniques for monitoring the spread of the CO_2 plume, assessing the area of high pressures caused by the CO_2 injection, and determining that the CO_2 plume is migrating into zones that do not damage resources or jeopardize the integrity of the reservoir [43]. Besides this, the plume of CO_2 should be monitored continuously within the reservoir to ensure that freshwater aquifers and ecosystems are well protected.

Monitoring, verification and accounting plans are necessarily related to risk analysis and subsequent mitigation measures. The expected range values of the different parameters associated with the performance of a CGS can be predicted by monitoring, which supposes an important step forward to an appropriate safety and risk analysis. At the same time, risk analysis allows the identification of the most important elements affecting the behavior of the CO_2 storage system. The visualization of these elements is of great interest in order to avoid mitigation or corrective measures. Consequently, it has to be analyzed the possible leakage pathways that threat the safety of the CO_2 storage facility, also considering the existing and novel mitigation tools and/or remediation measures in case of CO_2 escapes from a CGS. These techniques can be applied whenever the performance of the CO_2 storage system is not as the originally expected. Mitigation methodologies and mitigations tools are dealt with in Sect. 5.2.

5.1 Methodology for the Measurement of CO_2 Leakages and Dissolved and Free Associated Gases

One of the most important aspects concerning the performance assessment of a CGS is to increase the knowledge of the interaction between CO_2 and the storage and sealing formations, as well as the physico-mechanical resistance of the cap rock. Measurements to be carried out in a CGS constitute important tools to evaluate the capacity of the sealing formation or cap rock to retain CO_2, as well as dissolved and free associated gases. Consequently, CO_2 leakages and associated gases either dissolved or free, could indicate that the integrity of the CGS is jeopardized. For this reason, monitoring of these gases through their measurement should be carried out periodically in order to assess: (i) the performance of the CO_2 storage system; (ii) the capacity of the sealing formation to retain these gases; and (iii) the possible impacts of these gases released on the environment and people.

These measurements mainly include CO_2, either dissolved or free, diffuse soil CO_2 flux and CO_2 contents in soils (~1 m depth). Nevertheless, among the dissolved and free gases the concentration of N_2, O_2, Ar, CH_4, Ne, He, H_2 and ^{222}Rn could be also determined. If possible, it is also advisable to determine the concentration of ^{222}Rn in soils (~1 m depth) since this radioactive gas has frequently been used for the detection of fracture/fault systems that constitute potential pathways for gas leakages [44, 45]. Recently, ^{222}Rn determinations have been also used for monitoring the migration of CO_2 from a CGS, since CO_2 acts as a carrier gas for ^{222}Rn in a regime of advective transport and, consequently, CO_2 escapes from deep-seated sources may carry significant amounts of ^{222}Rn [46–49]. Therefore the determination of ^{222}Rn can also be indicative, in an indirect way, of the CO_2 escapes from a CO_2 storage system.

5.1.1 Methods of Sampling and Analysis of Dissolved Gases

The methods for determining the composition of dissolved gases (CO_2, N_2, O_2, Ar, CH_4, Ne, He, H_2) are generally conducted following two different criteria: (i) the total extraction; and (ii) the partial extraction. The method based on the first criterion [50] uses mechanical pumps and it is rarely used since it is complex and long sampling times are required. Furthermore, the total extraction of gases is not verified.

The partial extraction process generally involves the use of an inert gas (Ar, He, N_2) as carrier [51–53]. The carrier gas is introduced into the sample holder containing the liquid (Fig. 9a) therefore causing the partial extraction of the dissolved gases. This method, although it is characterized by its speed of execution and the availability of the materials, has the following disadvantages: (i) the concentration of the carrier gas cannot be determined; (ii) the injection of the carrier gas is complex and involves a high risk of contamination of the water sample; and (iii) the quantity of the gas extracted is generally low due to the dominant presence of the carrier gas. Consequently, these drawbacks limit the applicability of this method, often restricted to the determination of few species [54].

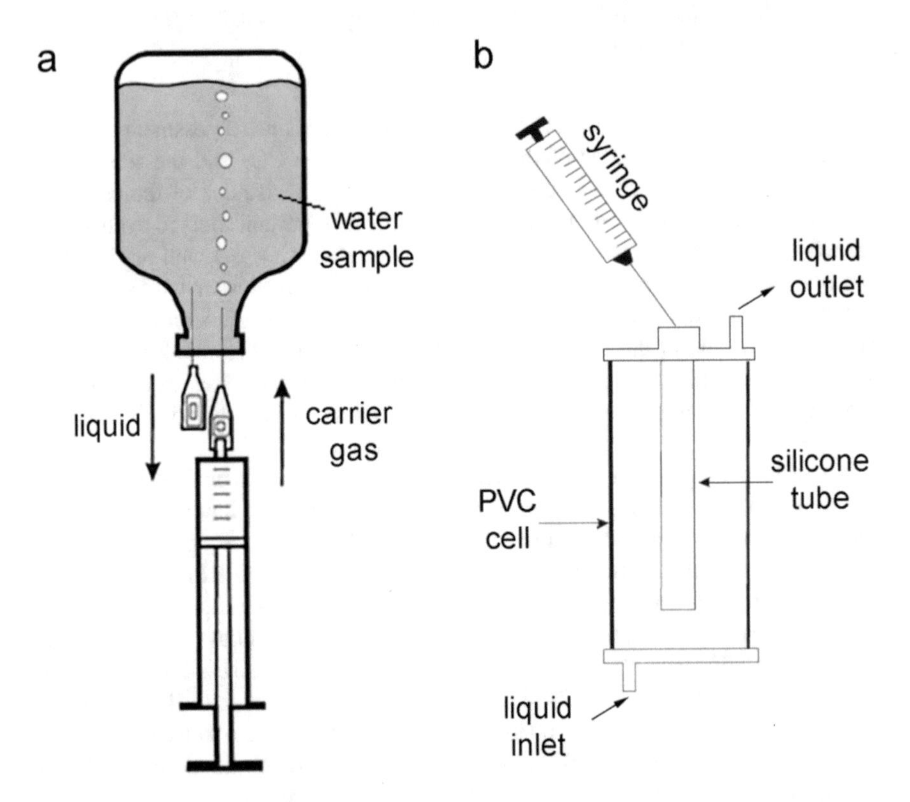

Fig. 9 Systems used for the extraction of dissolved gases: **a** by introducing a carrier gas, **b** through a permeable membrane ([54] modified)

Another method for the extraction of dissolved gases is by means of a silicone tube located inside a PVC cell, in which a constant flow of the liquid is maintained in order to extract the volatile compounds (Fig. 9b) [55, 56]. Nevertheless, this technique is difficult to use directly in the sampling site since it requires large amounts of sample. Furthermore, the permeation process through the silicone tube can cause fractionation of the gases, consequently modifying their original composition [54].

In addition to the abovementioned methods, dissolved gases can also be sampled by means of: (i) Niskin bottles, which are appropriate to collect water samples within the water column at different depths; (ii) the glass syringe method; and (iii) the direct immersion of vials of ~200–300 mL in which the vacuum (10^{-1}–10^{-2} Pa) is previously formed [54].

Sampling and analysis for ^{222}Rn dissolved is different with respect to the previous methods since samples are collected in low diffusion vials and filled up to the half with the so-called "scintillation cocktail". Spectra of ^{222}Rn and its descendants allow calculating the concentration of ^{222}Rn, expressed in Bq/L, as well as its uncertainty. Finally, the concentration of ^{222}Rn at the time of sampling was obtained considering its half-life (3.8 days). The wide popularity of liquid scintillation counting (LSC) is a consequence of numerous advantages, which are high efficiencies of detection, improvements in sample preparation techniques, automation including computer data processing, and the spectrometer capability of liquid scintillation analyzers permitting the simultaneous assay of different radionuclides. However, the main drawback of LSC is one of sensitivity.

5.1.2 Methods of Sampling and Analysis of Free Gases

The method used for sampling free gases is different depending if water sample shows or not bubbling. For the first case, the method basically consists of covering the wellhead with a latex bag (e.g. swimming cap) and then waiting for a "gas bag" (Fig. 10a). The gas is subsequently extracted (Fig. 10b) and, finally, it is injected into a vial previously filled with distilled water and punctured with a double-wall entry needles (Fig. 10c). The gas injection displaces the water through the aforementioned needle, accumulating this gas inside the vial. This method is quick, economic and easy to apply, although it is conditioned to the presence of bubbling waters.

For the second case, when no bubbling waters appear, it has to be firstly checked the presence of CO_2 by means of a portable CO_2 IR detector either at the wellhead or at depth. Once CO_2 is detected, the method consists of pumping this gas through a membrane pump through a tube, which has to be located at the depth in which the gas is detected. The output of the pump is connected to another tube, which in turn is attached to a hypodermic needle (Fig. 11). The gas transfer to the vial is performed following the same abovementioned method. This method is slower and more expensive compared to the previous one. Although can be tedious in operation, it has the main advantage that it can be applied in most of the wells.

Chemical determination of free gases can be carried out by means of a gas chromatograph coupled to a DSQ quadrupole mass spectrometer.

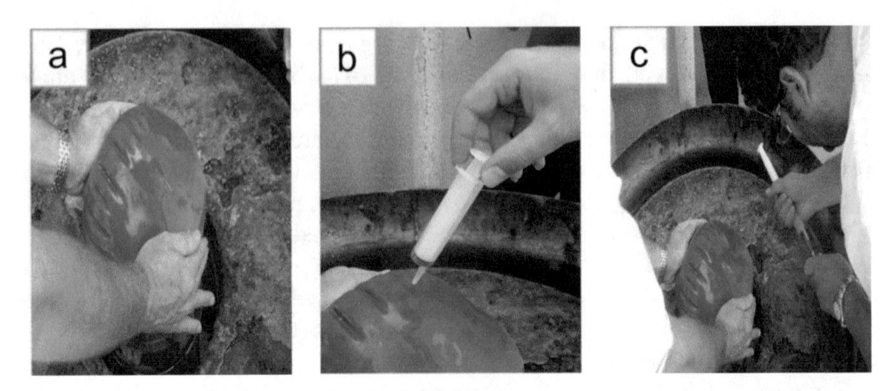

Fig. 10 **a** Latex bag used to retain gases in the wellhead for bubbling waters. **b** Plastic syringe with a hypodermic needle attached to extract the gas. **c** Injection of the gas into the vial filled with distilled water

Fig. 11 Example of a free gas sampling in a non-artesian well, once CO_2 was previously detected

5.1.3 Methods of Sampling and Analysis of Surficial CO_2 Flux

In relation to surficial CO_2 flux, measurements should be performed under favorable weather conditions, particularly during dry and meteorologically stable periods, in order to avoid the possible influence of rainfalls and the subsequent soil humidity. Since CO_2 is relatively soluble in water, environmental conditions are very important since they considerably affect their corresponding values.

Diffuse CO_2 flux was measured through the accumulation chamber method [57–63], which was originally used for agriculture purposes [59–62]. However, this method has extended its applications in the last two decades, including the measurements of CO_2 degassing in volcanic and geothermal environments [64–70] and for monitoring emissions from landfills [71, 72] being the main advantages its sensitivity, low cost, simple operability and high-speed data acquisition. On the contrary, the main drawbacks of this method is that diffuse CO_2 flux measurements can be affected by different factors, such as the variability of the surficial parameters (porosity, permeability), biological respiration, meteorological parameters (temperature, atmospheric pressure, wind speed), etc.

The material used for the diffuse CO_2 flux measurements includes: (i) an inverted chamber, with known dimensions, composed by a device that mixes the air in the chamber headspace; (ii) an Infra-Red (IR) spectrophotometer; (iii) an Analogical–Digital (AD) converter; and (iv) a Palmtop Computer (PC) (Fig. 12). To perform these measurements, the accumulation chamber is placed above the soil surface, allowing the CO_2 accumulation. Then, the gas is pumped towards the CO_2 IR detector with a flow rate of ~20 mL s^{-1}. Later, the gas is returned to the camera, therefore minimizing the disturbances of the gas naturally released from soil. Finally, the signal emitted by the IR is transmitted by the AD to the PC. In order to convert the volumetric

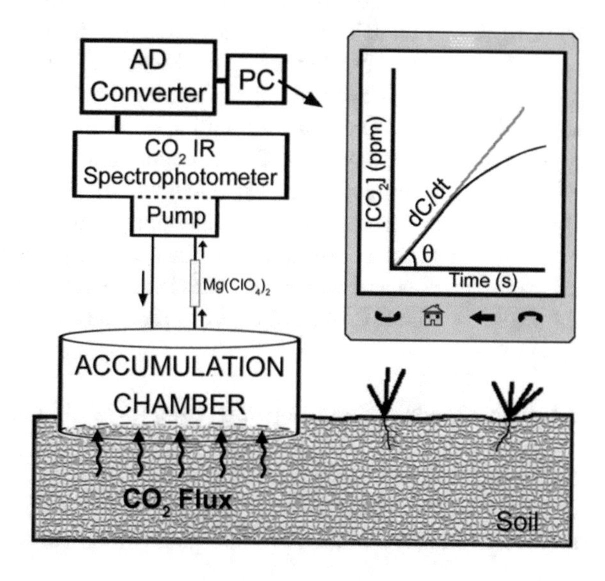

Fig. 12 Schematic representation of the accumulation chamber method used for the diffuse soil CO_2 flux measurements [48]

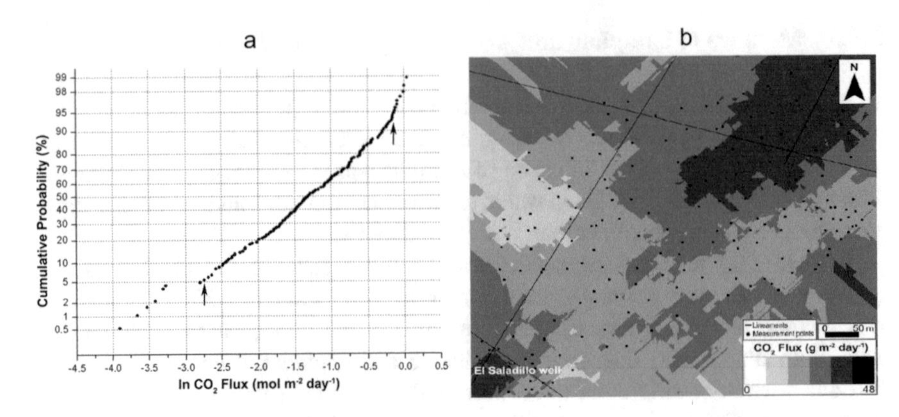

Fig. 13 **a** Cumulative probability plot corresponding to ln CO_2 flux showing the existence of three populations by means of the two inflection points, identified by arrows. **b** Mapping of the surficial distribution of the diffuse soil CO_2 flux by means of kriging estimation. Both examples are from the El Saladillo site (Murcia, SE Spain) [48]

concentration obtained (ppm s^{-1}) into mass concentration units (g m^{-2} day^{-1} or mol m^{-2} day^{-1}), it has to be considered the temperature, pressure and the volume of the chamber [73].

Computation of the total CO_2 flux is performed according to Sinclair [74] method, which is a graphical procedure usually used for geochemical data consisting of grouping the CO_2 values in different log-normal populations by considering the inflection points. Consequently, this method uses probability graphs, being a single log-normal population represented by a straight line, whilst a curve with $n - 1$ inflection points shows the theoretical distribution of n overlapped log-normal populations (Fig. 13a). Therefore, different populations from a data set can be recognized by using this method. The parameters needed to determine the total CO_2 flux of each population are calculated by using the Sichel [75] method, including the estimated percentage of each observed population, the flux mean value and the corresponding standard deviation. The total CO_2 output for each population is calculated by multiplying the site area, the ratio of each population and the mean CO_2 flux value. The 95% confidence interval was also calculated by using the Sichel's t-estimator [75]. By adding the sum of each individual population, it can be obtained the total CO_2 released to the surface. Besides this, these data can also be processed by means of kriging estimation and sequential Gaussian simulation methods [76], in order to map the spatial distribution of the CO_2 flux (Fig. 13b).

5.1.4 Methods of Sampling and Analysis of CO_2 and ^{222}Rn in Soils

Similarly to surficial CO_2 flux, CO_2 and ^{222}Rn concentration (~1 m depth) should be measured during dry and meteorologically stable periods, since these gases are relatively soluble in water and consequently their concentrations could be modified.

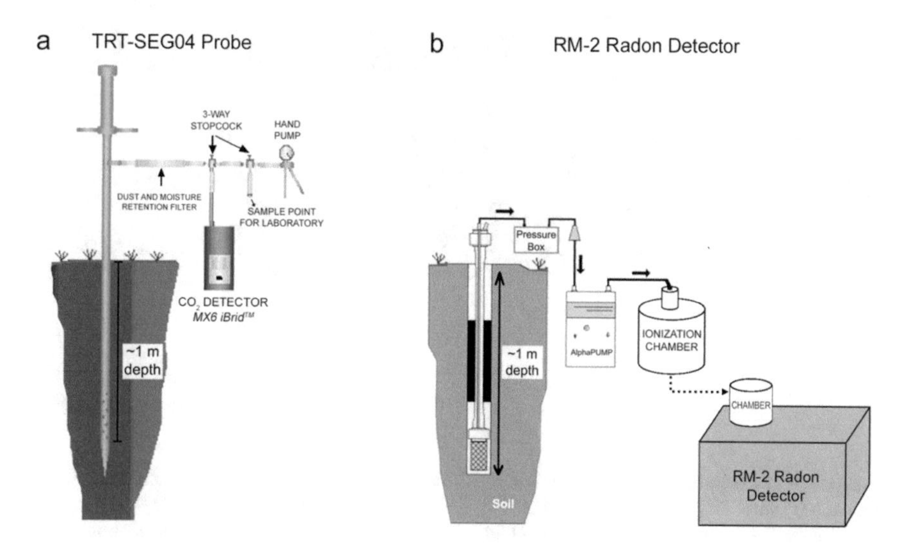

Fig. 14 **a** TRT-SEG04 probe for CO_2 concentration measurements. **b** RM-2 radon detector for determining ^{222}Rn concentration [49]

For CO_2 measurements, a probe for extracting soil gases is used, pumping the soil gases to a CO_2 detector. If necessary, a hand pump can be additionally coupled to extract the gases (Fig. 14a). For ^{222}Rn concentration, a Radon detector is used, being composed of an air suction pump coupled to ionization chambers equipped with a counter-photomultiplier device (Fig. 14b). The main advantages of both methods are their simplicity and that they are designed for in situ rapid analysis of CO_2 and ^{222}Rn, while the most important drawback is that measurements can be affected by the physical characteristics of the soil, especially porosity and moisture content, because they affect the gas transport in the soil.

In addition, it is essential to compile a base map of the emissions of free gases before the CO_2 injection, which can be used as a reference to compare it to others that will be carry out after the CO_2 injection at the site selected for CO_2 storage. An increase in the concentration of both gases in soils could be indicative of failures in the cap rock of the CGS. Therefore, the main objective is the detection, sampling, measurement and characterization of dissolved and free gases of the site selected, in order to determine variations in the concentrations once the anthropogenic CO_2 has been injected. The isotopic signature of the CO_2 detected in surface, either as dissolved inorganic carbon (DIC) or as a free gas, can serve as a tracer of the CO_2 stored.

5.1.5 Isotopes

The isotopic characterization of the dissolved and free gases is useful to determine their origin. Particularly, the isotopic values of C are used to determine the source

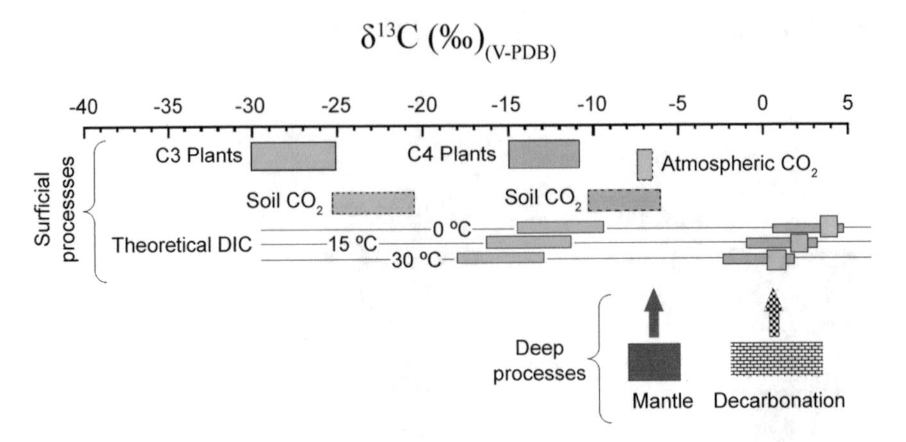

Fig. 15 Representation of the theoretical $\delta^{13}C$ values from the possible carbon sources, which include: (i) surficial processes: C3 plants [79], C4 plants [80], and atmospheric CO_2 [81]; and (ii) deep processes (inorganic C): mantle [82] and decarbonation [83]. CO_2 in the soil is about 4.5‰ heavier than the plant biomass [84, 85]. The isotopic difference between CO_2 and DIC depends on both pH and temperature. This value is close to 0‰ at about pH 5, but is relatively independent of pH between 7.5 and 8 [86]. For the theoretical DIC calculation, the calcite-CO_2 equation described by Romanek et al. [86] for temperatures of 0, 15 and 30 °C has been considered (figure modified after [87])

of CO_2 and therefore can represent an excellent tracer for CO_2 [77] considering the very negative $\delta^{13}C$-CO_2 values (~−30‰) related to the fossil fuel combustion [78]. In order to establish the possible carbon sources, the theoretical $\delta^{13}C$ values for the main carbon reservoirs can be plotted (Fig. 15), by including: (i) surficial processes: C3 plants [79], C4 plants [80], and atmospheric CO_2 [81]; and (ii) deep processes (inorganic C): mantle [82] and decarbonation [83].

In addition to $\delta^{13}C$, there are several isotopes than can be used to support the origin of gases. Among them, it can be highlighted the isotopes of noble gases, particularly He and Ne, which are typical trace gases of natural CO_2 reservoirs that can be used to differentiate inorganic sources.

Helium is highly diffusive with a diffusion coefficient about ten times that of CO_2 [88] also being physically stable, chemically inert and non-biogenic. Moreover, the $^3He/^4He$ ratio can be used to trace the presence of mantle magmas and deep gases, so it is frequently applied to distinguish between mantle and cortical sources, since mantelic CO_2 tends to be ^3He-enriched [89]. The ratio R/Ra (where R is the measured $^3He/^4He$ ratio and Ra is that of the air, i.e. 1.39E−06) can be as low as 0.0001 in the crust due to the radioactivity of U and Th and the formation of α particles (^4He), although this ratio is usually around 0.02 from crustal fluids [90–92]. Nevertheless, R/Ra ratio can take different range values in other geodynamic environments, such as: ~8 ± 1 in the Mid-Ocean Ridge Basalt (MORB) coming from the upper mantle [93–96]; ~10–30 in the Ocean Island Basalt (OIB) indicating a helium degassing source from the lower mantle [97, 98]; and ~5–8, related to subduction zones [90]. On the other hand, Neon is a light and very inert atmospheric gas with a $^4He/^{20}Ne$ ratio of 0.318 in air and 0.274 in waters [99]. Nevertheless, this ratio is $1 \cdot 10^7$ for cortical fluids and 1000 for mantle fluids [90]. Consequently, relatively low values of the $^4He/^{20}Ne$ ratio indicate that the sample has an important atmospheric contribution.

Finally, $\delta^{15}N$ values are widely applied to trace volatile sources in hydrothermal or volcanic systems. For this reason, they are usually used to investigate mantle geochemistry and global volatile cycles. According to the origin of the sample, this nitrogen isotope can take different values: (i) 0‰, related to the atmosphere; (ii) -5 ± 2‰, which is assigned to the upper mantle; and (iii) 0–10‰, derived from subduction zones sediments [100, 101].

All these stable isotope ratio analysis can be determined by using a mass spectrometer, being very convenient to follow the referencing strategies and techniques described by Werner and Brand [102].

5.2 *Mitigation Tools*

It is well known the existence of a wide variety of methods for mitigating and/or correcting the possible effects of CO_2 leakages from a CGS. It has also been demonstrated that the mitigation or correction techniques are more effective close to the source of the CO_2 escape rather than near the surface, where the detection of CO_2 is more difficult since it tends to be dispersed.

Undesired CO_2 leakages could occur within or out of the reservoir via faults/fractures or along the wellbore, being three the main causes of the loos of the safe behavior of the CO_2 storage complex [103]: (i) the loss of the reservoir's integrity; (ii) the existence of fractures and/or faults that could constitute possible pathways for CO_2 leakages; and (iii) the loss of the well integrity. These causes, as well as their possible mitigation/correction measures, are discussed in detail in the following sections. Firstly, it should be remembered that the application of these measures is the last option to consider since an adequate previous planning, including monitoring and risk assessment, could avoid carrying out these unexpected measures.

5.2.1 Loss of the Reservoir's Integrity

The loss of the reservoir's integrity can be mainly due to the following different reasons: (i) a discontinuity or compartmentalization of the geological storage formation, therefore leading to a significant increase of the pressure in the injection well; (ii) an unexpected fluid flow within the reservoir, e.g. the spread of the CO_2 plume beyond the desired region, such as a fault/fracture zone or discharge point, or the migration of the CO_2 plume through the cap rock; and (iii) the creation or reactivation of faults and/or fractures in the reservoir, or in the cap rock, caused by stress changes during CO_2 injection [104–107], since the stress path has a deep effect on stress dynamics and fracturing/faulting when injecting into a depleted reservoir [105].

Corrective measures, basically based on pressure, can be applied within the CO_2 reservoir [103, 108]. These measures include: (i) the permeability reduction by injecting gels/foams or by immobilizing the CO_2 through solid reaction products [109]; (ii) the change of injection strategy, which can potentially prevent or retard CO_2 from arriving at undesired migration pathways (faults, fracture zones or discharge points) and could also represent an efficient measure compared to active remediation from an economic point of view; and (iii) the localized injection of brine, hence creating a competitive fluid movement.

The methods aimed to reduce the permeability of CO_2 storage reservoirs by using the polymer-gel injection are conditioned by different parameters such as polymer type, molecular weight, polymer concentration, crosslinker concentration, ratio of polymer-to-crosslinker and temperature [108].

Fluid movement within a CO_2 reservoir is based not only on reservoir properties (structural dip or spatial heterogeneity in permeability and/or porosity) but also it can be managed by distributing the reservoir pressure. In this sense, CO_2 migration can also be managed by either brine extraction or CO_2 backproduction [110]. In any case, these both measures create pressure gradients towards the extraction point, consequently enforcing a specific flow direction [103].

5.2.2 Existence of Fractures and/or Faults

The possibility of reducing or disrupting CO_2 leakages through faults and/or fractures has been considered by assessing the efficacy of reducing pressure to lower the leakage rate or by using sealants (e.g. gels or foams) to interrupt the escape. In addition, other possibilities have been tested, like transferring CO_2 through a fault in a compartment originally unconnected to the main reservoir, improving the sealing capacity of the cap rock by injecting N_2 before or during CO_2 injection [111]; or by generating a flow barrier above the cap rock by creating a reverse pressure gradient.

Remediation of CO_2 leakages by CO_2 flow diversion

The principle of remediation of CO_2 leakages by CO_2 flow diversion towards close compartments from the CO_2 storage reservoir through hydraulic fractures or deviated wells (Figs. 16 and 17) requires the creation of a pathway for fluid migration between the CO_2 storage reservoir and the leaky and neighboring compartments, since the CO_2 reservoir and neighboring compartments are originally not connected (see Fig. 17). In this sense, compartmentalized saline aquifers or gas reservoirs represent geological settings potentially suitable for remediation by flow diversion.

In the case of relevant CO_2 leakages from a CGS, pressure relief can be achieved by diverting CO_2 from the CO_2 storage complex to non-connected parts of the reservoir, or to adjacent aquifers and/or reservoirs. This fluid migration can be performed by hydraulic fracturing (fracking) across a sealing fault that separates adjacent compartments, or also by drilling a well. The effects of flow diversion as a remediation option were evaluated from a real field case in the North Sea, concluding that this flow is a possible remediation option for specific depleted gas fields or saline aquifers, being

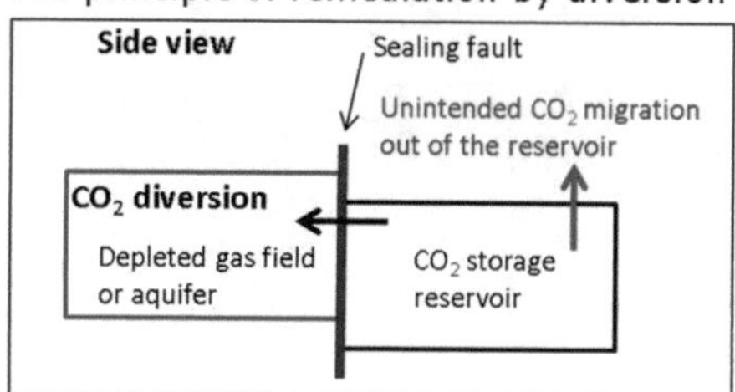

Fig. 16 The principle of remediation of CO_2 leakages by flow deviation from the CO_2 storage reservoir to the adjacent unconnected reservoir. Hydraulic connection between the two reservoirs separated by a sealing fault could be achieved by drilling a deviated well or by creating hydraulic fractures through the fault seal ([112] modified)

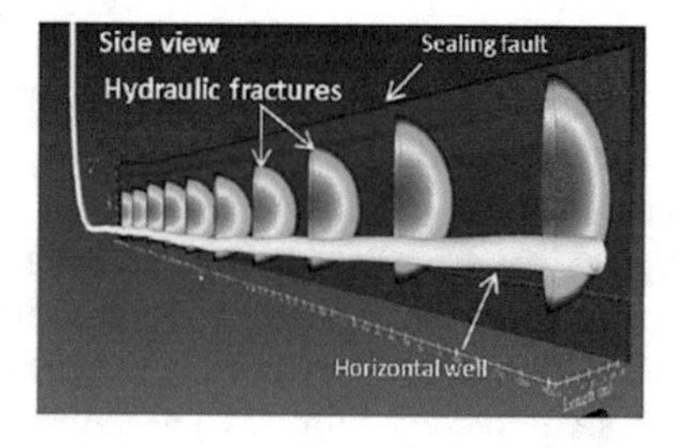

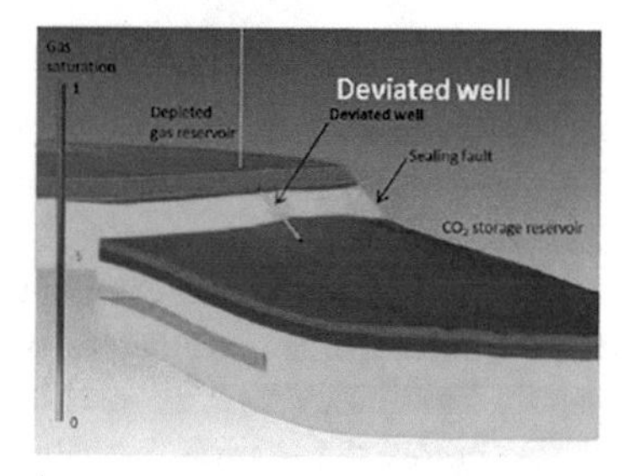

Fig. 17 a Breaching of fault seal by hydraulic fracturing, b or by drilling a deviated well. These two methods enables lateral migration of fluids between the two adjacent reservoirs separated by a sealing fault [112]

two the key factors controlling the efficiency of flow diversion: (i) the conductivity and the pressure difference between the two reservoirs; and (ii) the permeability of the receiving reservoir. This type of remediation in a saline aquifer is relatively slow compared to an adjacent depleted gas field, due to the small pressure difference between the two compartments [113].

Fault sealants

The oil and gas (O&G) industry generally uses different techniques to reduce the flow rate of a given fluid or to maximize oil or gas recovery by injecting fluids with specific properties. Some of these methods should be appropriately selected or adapted for reducing or interrupting CO_2 escapes through fractures and/or faults, such as the injection of polymer-gel in order to seal the fault, consequently diverting the flow within the reservoir [114].

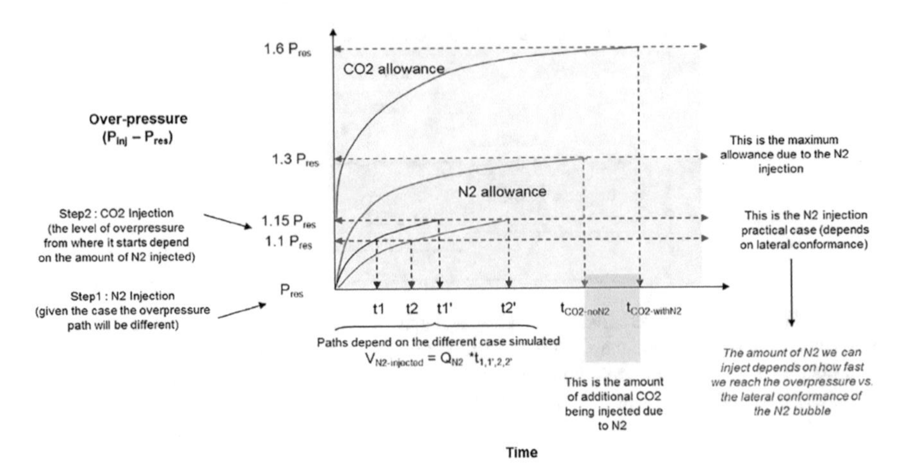

Fig. 18 Conceptual design of a N_2 injection previously to the injection of CO_2 [111]

Barriers

The use of barriers is mainly focused on: (i) checking a mitigation way to prevent CO_2 leakages by injecting N_2 in the cap rock; and (ii) testing a hydraulic barrier after CO_2 leakage by injecting water in a permeable layer above the cap rock.

Regarding the first use, current CGS projects in deep saline aquifers are naturally limited, among other parameters, by entry pressures encountered in cap rocks, consequently limiting over-pressures allowed during the storage process. The injection of N_2 just below the sealing formation, previously to the injection of CO_2, could be a protective measure to increase the storage safety by lowering the leakage risk and by increasing the maximum allowable reservoir pressure [111]. The concept governing the injection of N_2 is summarized in Fig. 18.

The concept of the beneficial impact of the injection of N_2 on the cap rock, consequently increasing the pressure, is based on the higher N_2–brine interfacial tension (IFT) compared to the CO_2–brine IFT. As a maximum possible effect (i.e. pure N_2-brine systems) IFT could increase by two times, yielding correspondingly to the same increase of allowable pressure. Nevertheless, the main disadvantage is that N_2 injection decreases the CO_2 storage capacity and the trade-off must be analyzed carefully, since the IFT spread decreases rapidly with the mixing ratio of CO_2 in the N_2 [111].

As regards the second use, this corrective measure aims at countering the main driving force of the CO_2 upwards migration which is the pressure build-up under the leak by injecting brine into the shallower aquifer, thus creating a hydraulic barrier [115]. This remediation technique, which can be applied at low cost but is only temporary, will decrease the CO_2 leakage rate occurring across the cap rock.

5.2.3 Loss of Well Integrity

Measures aimed to mitigate or correct the loss of well integrity in case of CO_2 escapes are well documented and, consequently, can be consulted at the best-practice recommendations from the O&G industry, which has a great experience in this field. Therefore, this best-practice portfolio of remediation technologies can also be applicable to CO_2 injection wells. Furthermore, new developments and emerging technologies should also be considered, including gels, smart cement and polymer resins.

Oil and Gas best practices

Experience from O&G industry has revealed that wells constitute the highest risk of CO_2 leakages from a CGS [116] being mainly caused by the failure of the barrier elements. Carbon dioxide leakages mainly occur due to the poorly cemented casing, casing failure and/or improper abandonment [117].

From the best-practice recommendations of the O&G industry, a generic and systematic approach has been used to discuss the most critical well barrier elements, only considering one type of well and a typical CO_2 injection well equipped with primary and secondary well barrier elements (Fig. 19). The best analogue for a CO_2 injection well in an O&G setting has been employed for this analysis, which could be considered as an operating gas well with high CO_2 contents and high gas/oil ratio. The basic well design for both O&G and CO_2 wells is almost identical except for the materials used, which are more critical for CO_2 [112].

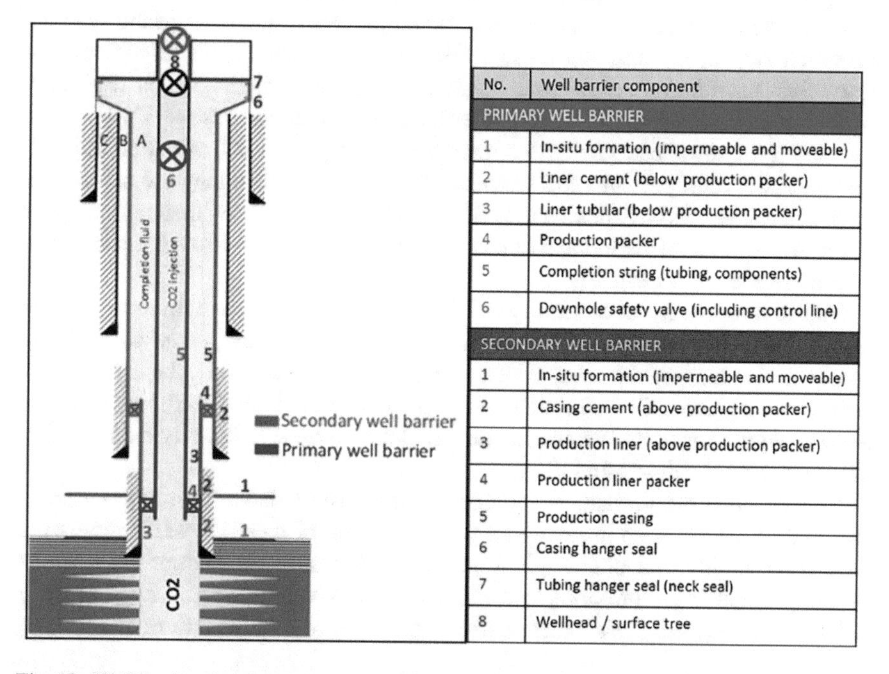

No.	Well barrier component
PRIMARY WELL BARRIER	
1	In-situ formation (impermeable and moveable)
2	Liner cement (below production packer)
3	Liner tubular (below production packer)
4	Production packer
5	Completion string (tubing, components)
6	Downhole safety valve (including control line)
SECONDARY WELL BARRIER	
1	In-situ formation (impermeable and moveable)
2	Casing cement (above production packer)
3	Production liner (above production packer)
4	Production liner packer
5	Production casing
6	Casing hanger seal
7	Tubing hanger seal (neck seal)
8	Wellhead / surface tree

Fig. 19 Well barrier elements for a typical CO_2 injection well [112]

The occurrence of CO_2 leakages means that both primary and secondary barrier components of the CO_2 injection well fail simultaneously. Early escape events are mainly often related to: (i) an inaccurate well design; (ii) an incorrect material selection; and (iii) a wrong installation of the well barrier elements. However, late leakages are frequently associated with: (i) corrosion and/or erosion of materials; and (ii) degradation and/or fatigue of materials. Finally, failures or defects in the well barrier elements are linked to common mitigation and remediation techniques generally used in the O&G industry. Since these practices can be complex and expensive, it is advisable to perform some preventive actions in order to reduce the risk of failures [112].

Novel materials and emerging technologies

New developments and breakthrough technologies for mitigation and remediation of CO_2 leakages from wells are being tested nowadays. The objective is to inject a solution in the surroundings of a well in a selected depth interval (usually a few meters) in order to seal the near well bore formation, therefore reducing porosity and permeability ideally down to zero and not allowing CO_2 to flow at that depth. Consequently, the porosity should be filled with a solid, being this solid the result of the precipitation of some components of the injected solution [118]. Although there is a wide variety of methods that can be used to treat the surroundings of a well [119] there are new emerging processes that are promising, such as the use of: (i) CO_2 reactive suspensions; (ii) polymer-based gels; (iii) smart cements with a latex-based component [120]; and (iv) polymer resin for squeezing.

CO_2 reactive suspensions have been studied for reducing the permeability in the near-well region, highlighting those suspensions that use silicate based solutions since they have high performance, long term chemical stability, good injectivity (low viscosity and no particles) and no or little environmental impact [118].

The use of smart cements with a latex-based component is focused on the self-sealing under high pressure and temperature conditions when they are exposed to CO_2. These cements have demonstrated to be effective in reducing their permeability either through the casing-cement or cement-rock interfaces, or through the fractures within the cement itself [120].

The sealing ability of a commercially available temperature-activated polymer resin with respect to cement failure at laboratory scale was proved to be fairly successful in plugging the designed leak paths for the two selected leakage scenarios: cement-casing debonding and fractures in annular cement. The results showed that the permeability and the average fracture thickness were significantly reduced after the treatment with this resin [121].

Finally, once the mitigation and corrective measures have been exposed, it is essential to always keep in mind that the application of these measures represents the last possibility to avoid CO_2 leakages. For this reason, an appropriate previous planning, including monitoring and risk analysis, is very important and useful in order to not carry out these undesirable measures that reveal the failure of a CGS.

6 Concluding Remarks

The proposed methodology assumes an approach to the problems of risk analysis derived from CGS activities. The development of models based on BN for the description of these systems is not an easy task. However, although very sophisticated methods are actually applied, it is an attractive tool because it allows the possibility of making decisions under conditions of uncertainty together with the fact of being a natural way of making connections between the different elements and the simplicity of its maintenance. Furthermore, the proposed methodology, given its conceptual development, allows realizing mathematical analyzes (zones of maximum and minimum variation, zones of stability, etc.), sensitivity analysis to determine both the variables that contribute the most uncertainties to the system as well as the different conceptual models, which are fundamental for the treatment of system uncertainties, etc., all of which are basic activities in the analysis of risks of any CGS project.

From the development of the proposed methodology and its application to a study area, it can be concluded that it allows evaluating the probability of risk of leakage probability from an area with potential capabilities as a CGS site, solely from qualitative-type of data. From the comparison of the proposed methodology with the methodology of recognized prestige called Selection and Classification of Formations it can be concluded that they coincide in the qualification of the area. Both evaluations have a qualitative character. However, although the route of the Selection and Classification of Formations methodology ends at this point, for the proposed methodology it means the starting point, since, starting from the relationships already established between the different variables, it will gradually progress to quantitative modeling.

The BN formalism allows generating a Risk Analysis process in which progressively and without a solution of continuity, it would pass from being based on modeling of a pure qualitative type, to Risk Analysis based on qualitative-quantitative mix of modelling to, finally, attain a RA based on pure quantitative modeling. This would allow to embrace the CGS project as a whole, through a continuous RA process, from the initial stages, characterized by a shortage of available information, thanks to the adoption of a subjective perspective of the probability concept, and to the application of EJ. Undoubtedly, these initial analyses will not be without biases and heuristics. However, this initial problem would be progressively overcome based on the advance in the available information and the generation of modeling based on physical/chemical-mathematical models that would be gradually replace the qualitative estimates based on EJ [122–124].

Furthermore, this RA should help identify not only potential locations for CGS sites, but also approximations for enhanced measurement, monitoring and verification activities. Monitoring is an essential part of the entire risk management for CGS, as well as the remediation measures to be applied in case of unexpected events that can compromise the safety of a geological storage of CO_2.

Acknowledgements The studies showed in this chapter has been funded by Fundación Ciudad de la Energía (Spanish Government) (https://www.ciuden.es), by the European Union through the Compostilla OXYCFB300 project and in the framework of the Project entitled: "Tecnologías avanzadas de generación, captura y almacenamiento de CO2 (PSE-CO2)", supported by the former Spanish Ministry of Science and Innovation and the EU FEDER funds under award number PSE-120000-2008-6.

This document reflects only the authors' view and that European Commission or Spanish Government are not liable for any use that may be made of the information contained therein.

Glossary

BN Bayesian Networks
CCS Carbon Capture and Storage
CGS CO_2 geological storage
DIC Dissolved inorganic carbon
EJ Expert Judgment
FEP Features, Events and Processes
FMEA Failure Mode and Effect Analysis
HSE Health, Safety and the Environment
IFT Interfacial tension
IR Infra-Red
KPI Key Performance Indicators
LSC Liquid scintillation counting
MCA Multi-Criteria Assessment
MORB Mid-Ocean Ridge Basalt
OIB Ocean Island Basalt
RM Risk management
SRF Screening and Ranking Framework
SWIFT Structured "What-If" Techniques
THMQ Thermo-hydro-mechanical-chemical
VEF Vulnerability Evaluation Framework

References

1. Bachu, S. (2008). CO_2 storage in geological media: Role, means, status and barriers to deployment. *Progress in Energy and Combustion Science, 34,* 254–273.
2. European Community. (2012). Implementation of directive 2009/31/EC on the geological storage of carbon dioxide. In *Guidance document 1: CO_2storage life cycle risk management framework 2012-06-15*. https://op.europa.eu/s/oayY
3. van Egmond, B. (2006). *Developing a method to screen and rank geological CO_2storage sites on the risk of leakage* (NWS-E-2006-108). Copernicus Institute, Department of Science, Technology and Society.

4. Pérez-Estaún, A., Gómez, M., & Carrera, J. (2009). El almacenamiento geológico de CO_2, una de las soluciones al efecto invernadero. *Enseñanza de las Ciencias de la Tierra, 17*(2), 179–189.
5. ISO. (2009). *ISO/IEC guide 73:2009. "Risk management-vocabulary"*. International Organization for Standardization.
6. PMI (Project Management Institute). (2000). *A guide to the project management body of knowledge (PMBOKR)* (2000 ed.). Newtown Square, PA: Project Management Institute.
7. IRGC. (2013). *Risk governance guidelines for unconventional gas development*. Lausanne: International Risk Governance Council. ISBN: 978-2-9700-772-8-2.
8. Condor, J., Unatrakarn, D., Asghari, K., et al. (2011). A comparative analysis of risk assessment methodologies for the geologic storage of carbon dioxide. *Energy Procedia, 4,* 4036–4043.
9. NETL—National Energy Technology Laboratory. (2017a). *BEST PRACTICES: Risk management and simulation for geologic storage projects* (Report DOE/NETL-2017/1846) (p. 114).
10. Li, Q, & Liu, G. (2017). Risk assessment of the geological storage of CO_2: A review. In V. Vishal & T. N. Singh (Eds.), *Geologic carbon sequestration.* Switzerland: Springer International Publishing. https://doi.org/10.1007/978-3-319-27019-7_13
11. SKI. (1996). *SKI Site-94: Deep repository performance assessment project* (SKI Report SKI 96:36). Swedish Nuclear Power Inspectorate, Stockholm, Sweden.
12. Stenhouse, M. J. (2001). *Application of systems analysis to the long-term storage of CO_2 in the Weyburn reservoir* (Monitor Scientific Report MSCI-2025-1). Denver, CO: Monitor Scientific LLC.
13. Garcia-Aristizabal, A., Kocot, J., Russo, R., et al. (2019). A probabilistic tool for multi-hazard risk analysis using a bow-tie approach: Application to environmental risk assessments for geo-resource development projects. *Acta Geophysica, 67,* 385–410.
14. Paraguassú, M. M., Câmara, G., Rocha, P. S., et al. (2015). An approach to assess risks of carbon geological storage technology. *International Journal of Global Warming 7*(1).
15. Pawar, R. J., Bromhal, G. S., Carey, J. W., et al. (2015). Recent advances in risk assessment and risk management of geologic CO_2 storage. *International Journal of Greenhouse Gas Control, 40,* 292–311.
16. Liu, W., & Ramirez, A. (2017). State of the art review of the environmental assessment and risks of underground geo-energy resources exploitation. *Renewable and Sustainable Energy Reviews, 76,* 628–644.
17. Hurtado, A., Recreo, F., Eguilior, S., et al. (2010). Aplicación de una evaluación preliminar de la seguridad y de los riesgos HSE a las potenciales ubicaciones de una planta piloto de almacenamiento geológico de CO_2. In *Comunicación Técnica en 10° Congreso Nacional del Medio Ambiente (CONAMA)*, Madrid, 22 al 26 de noviembre de 2010. ISBN: 978-84-614-6112-7.
18. Hurtado, A., Eguilior, S., & Recreo, F. (2014). Methodological development of a probabilistic model for CO_2 geological storage safety assessment. *International Journal of Energy and Environmental Engineering, 5*, 2–3.
19. Hurtado, A., Eguilior, S., & Recreo, F. (2015). *Modelo Probabilista de Evaluación Integrada del Comportamiento de la Planta de Desarrollo Tecnológico de Hontomín.* Versión 2 Informe Técnico CIEMAT 1346. Depósito Legal: M-26385-2011 ISSN: 1135-9420 NIPO: 721-15-005-6.
20. Pawar, R., Dilmore, R., Chu, S., et al. (2017). Informing geologic CO_2 storage site management decisions under uncertainty: Demonstration of NRAP's integrated assessment model (NRAP-IAM-CS) application. *Energy Procedia, 114,* 4330–4337.
21. Ruiz, C., Recreo, F., Prado, P., Campos, R., Pelayo, M., de la Losa, A., et al. (2007). Almacenamiento Geológico de CO_2. Criterios de selección de emplazamientos. *Informe Técnico CIEMAT, 1106*, 99 pp. ISSN: 1135-9420.
22. Bouc, O., & Fabriol, H. (2007). Towards a methodology to define safety criteria for CO_2 geological storage. In *Sixth Annual Conference on Carbon Capture & Sequestration*, Pittsburgh, PA, May 7–10, 2007.

23. Bouc, O., Audigane, P., Bellenfant, G., Fabriol, H., Gastine, M., Rohmer, J., & Seyedi, D. (2009). Determining safety criteria for CO_2 geological storage. *Energy Procedia, 1,* 2439–2446.
24. Stauffer, P. H., Viswanthan, H. S., Klasky, M. L., et al. (2006). CO_2-PENS: A CO_2 sequestration system model supporting risk-based decision. In *CMWR XVI*, Copenhagen.
25. Oldenburg, C. M. (2008). Screening and ranking framework for geologic CO_2 storage site selection on the basis of health, safety and environmental risk. *Environmental Geology, 54,* 1687–1694.
26. Toth, F. L. (Ed.). (2011). Geological disposal of carbon dioxide and radioactive waste: A comparative assessment. *Advances in Global Change Research, 44.* https://doi.org/10.1007/978-90-481-8712-6. International Atomic Energy Agency. ISBN: 978-90-481-8711-9.
27. Lanting, M., Hurtado, A., Eguilior, S., & Llamas, J. F. (2019). Forecasting concentrations of organic chemicals in the vadose zone caused by spills of hydraulic fracturing wastewater. *Science of the Total Environment, 696,* 133911.
28. Le Guenan, T., Gravaud, I., de Dios, C., Loubeau, L., Poletto, F., Eguilior, S., & Hurtado A. (2018). Determining performance indicators for linking monitoring results and risk assessment—Application to the CO_2 storage pilot of Hontomìn, Spain. In *Conference Paper. 14th Greenhouse Gas Control Technologies Conference (GHGT-14)*, Melbourne, October 21–26, 2018. Available at SSRN: https://ssrn.com/abstract=3366022
29. Pearl, J. (1998). *Probabilistic reasoning in intelligent systems: Networks of plausible inference.* San Mateo, CA: Morgan Kaufmann Publishers.
30. Xu, T., Apps, J. A., & Pruess, K. (2002). *Reactive geochemical transport simulation to study mineral trapping for CO_2disposal in deep saline arenaceous aquifers.* Lawrence Berkeley National Laboratory. También accesible en https://www.escholarship.org/uc/item/7hk8s1nx
31. Riaz, A., Hesse, M., Tchelepi, A., et al. (2006). Onset of convection in a gravitationally unstable diffusive boundary layer in porous media. *Journal of Fluid Mechanics, 548,* 87–111.
32. Michael, K., Bachu, S., Buschkuehle, B. E., et al. (2006). Comprehensive characterization of a potential site for CO_2 geological storage in Central Alberta, Canada. In *CO2SC Symposium* (pp. 134–138). Berkeley, CA: Lawrence Berkeley Laboratory.
33. Chang, K., Minkoff, S. E., & Bryant, S. L. (2008). *Modeling leakage through faults of CO_2stored in an aquifer* (SPE 115929, SPE ATCE). Denver, CO.
34. IPCC. (2005). In B. Metz, O. Davidson, H. de Coninck, M. Loos, & L. Meyer (Eds.) (431 pp.). Cambridge: Cambridge University Press.
35. Bowden, A. R., & Rigg, A. (2004). Assessing risk in CO_2 storage projects. *The APPEA Journal, 44,* 677–702.
36. ANCOLD (Australian National Committee on Large Dams). (2003). *'Life safety risks' in the ANCOLD guidelines on risk assessment.* ANCOLD.
37. INSHT. (2008). *Límites de exposición profesional para agentes químicos en España; Año 2008.* Instituto Nacional de Seguridad e Higiene en el Trabajo (INSHT).
38. U.S. EPA, 40 CFR Parts 144 and 146. (2008). *Federal requirements under the underground injection control program for carbon dioxide geologic sequestration wells. Proposed rule.*
39. Kovscek, A. R. (2002). Screening criteria for CO_2 storage in oil reservoirs. *Petroleum Science and Technology., 20*(7–8), 841–866.
40. Holtz, M. H. (2002). *Residual gas saturation to aquifer influx: A calculation method for 3-D computer reservoir model construction.* SPE Paper 75502. Presented at the SPE Gas Technologies Symposium, Calgary, Alberta, April.
41. Taber, J. J., Martin, F. D., & Seright, R. S. (1997). EOR screening criteria revisited—Part 1: Introduction to screening criteria and enhanced recovery field projects. *SPE Reservoir Engineering, 12*(3), 189–198.
42. Gozalpour, F., Ren, S. R., & Tohidi, B. (2005). CO_2 EOR and storage in oil reservoirs. *Oil & Gas Science and Technology—Revue IFP, 60*(3), 537.
43. NETL—National Energy Technology Laboratory. (2017b). *BEST PRACTICES: Monitoring, verification, and accounting (MVA) for geologic storage projects* (DOE/NETL-2017/1847) (88 pp.).

44. Ioannides, K., Papachristodoulou, C., Stamoulis, K., Karamanis, D., Pavlides, S., Chatzipetros, A., & Karakala, E. (2003). Soil gas radon: A tool for exploring active fault zones. *Applied Radiation and Isotopes, 59,* 205–213.
45. Walia, V., Lin, S.-J., Fu, C.-C., Yang, T. F., Hong, W.-L., Wen, K.-L., & Chen, C.-H. (2010). Soil-gas monitoring: A tool for fault delineation studies along Hsinhua Fault (Tainan), Southern Taiwan. *Applied Geochemistry, 25,* 602–607.
46. Rodrigo-Naharro, J. (2014). *El análogo natural de almacenamiento y escape de CO_2de la cuenca de Gañuelas-Mazarrón: implicaciones para el comportamiento y la seguridad de un almacenamiento de CO_2en estado supercrítico* (PhD thesis). Universidad Politécnica de Madrid, 427 pp. Published thesis. ISBN: 978-84-7834-738-4.
47. Rodrigo-Naharro, J., Nisi, B., Vaselli, O., et al. (2012). Measurements and relationships between CO_2 and 222Rn in a natural analogue for CO_2 storage and leakage: The Mazarrón Tertiary Basin (Murcia, Spain). *Geo-Temas, 13,* 1978–1981.
48. Rodrigo-Naharro, J., Nisi, B., Vaselli, O., et al. (2013). Diffuse soil CO_2 flux to assess the reliability of CO_2 storage in the Mazarrón-Gañuelas Tertiary Basin (Spain). *Fuel, 114,* 162–171.
49. Rodrigo-Naharro, J., Quindós, L. S., Clemente-Jul, C., et al. (2017). CO_2 degassing from a Spanish natural analogue for CO_2 storage and leakage: Implications on 222Rn mobility. *Applied Geochemistry, 84,* 297–305.
50. Carter, R., Kaufman, W. J., Orlob, G. T., & Todd, D. K. (1959). Helium as a ground water tracer. *Journal of Geophysical Research, 64,* 689–709.
51. Tonani, F. (1971). Concepts and techniques for the geochemical forecasting of volcanic eruption. In *The surveillance and prediction of volcanic activity* (pp. 145–166). Paris: UNESCO.
52. Sugisaki, R., & Taki, K. (1987). Simplified analyses of He, Ne and Ar dissolved in natural waters. *Geochemical Journal, 21,* 23–27.
53. Capasso, G., & Inguaggiato, S. (1998). A simple method for the determination of dissolved gases in natural waters. An application to thermal waters from Vulcano Island. *Applied Geochemistry, 13,* 631–642.
54. Tassi, F., Vaselli, O., Luchetti, G., et al. (2008). *Metodo per la determinazione dei gas disciolti in acque naturali* (Internal Report CNR-IGG, no. 2/2008). Florence.
55. Jacinthe, P. A., & Groffman, P. M. (2001). Silicone rubber sampler to measure dissolved gases in saturated soils and waters. *Soil Biology & Biochemistry, 33,* 907–912.
56. De Gregorio, S., Gurrieri, S., & Valenza, M. (2005). A PTFE membrane for the in situ extraction of dissolved gases in natural waters: Theory and applications. *Geochemistry, Geophysics, Geosystems, 6,* 1–13.
57. Baubron, J. C., Allard, P., & Toutain, J. P. (1990). Diffuse volcanic emissions of carbon dioxide from Vulcano Island, Italy. *Nature, 344,* 51–53.
58. Baubron, J. C., Allard, P., & Toutain, J. P. (1991). Gas hazard on Vulcano Island. *Nature, 350,* 26–27.
59. Chan, A. S. K., Prueger, J. H., & Parkin, T. B. (1998). Comparison of closed-chamber and bowen-ratio methods for determining methane flux from peatland surface. *Journal of Environmental Quality, 27,* 232–239.
60. Hutchinson, G. L., & Moiser, A. R. (1981). Improved soil cover method for field measurement of nitrous fluxes. *Soil Science Society of America Journal, 45,* 311–316.
61. Kanemasu, E. T., Power, W. L., & Sij, J. W. (1974). Field chamber measurements of CO_2 flux from soil surface. *Soil Science, 118,* 233–237.
62. Mitra, S., Jain, M. C., Kumar, S., et al. (1999). Effect of rice cultivation on methane emission. *Agriculture, Ecosystems & Environment, 73,* 177–183.
63. Parkinson, K. (1981). An improved method for measuring soil respiration in the field. *Journal of Applied Ecology, 18,* 221–228.
64. Bergfeld, D., Goff, F., & Janik, C. J. (2001). Elevated carbon dioxide flux at the Dixie valley geothermal field, Nevada; relations between surface phenomena and the geothermal reservoir. *Chemical Geology, 177,* 43–66.

65. Brombach, T., Hunziker, J. C., Chiodini, G., et al. (2001). Soil diffuse degassing and thermal energy fluxes from the southern Lakki plain, Nisyros (Greece). *Geophysical Research Letters, 28,* 69–72.
66. Chiodini, G., Cioni, R., Guidi, M., et al. (1998). Soil CO_2 flux measurements in volcanic and geothermal areas. *Applied Geochemistry, 13,* 543–552.
67. Chiodini, G., Frondini, F., Kerrick, D. M., et al. (1999). Quantification of deep CO_2 fluxes from Central Italy. Examples of carbon balance for regional aquifers and of soil diffuse degassing. *Chemical Geology, 159,* 205–222.
68. Chiodini, G., Frondini, F., Cardellini, C., et al. (2001). CO_2 degassing and energy release at Solfatara volcano, Campi Flegrei, Italy. *Journal of Geophysical Research: Solid Earth, 106,* 16213–16221.
69. Gerlach, T. M., Doukas, M. P., McGee, K. A., et al. (1998). Three-year decline of magmatic CO_2 emissions from soils of a mammoth mountain tree kill: Horseshoe Lake, CA, 1995–1997. *Geophysical Research Letters, 25,* 1947–1950.
70. Gerlach, T. M., Doukas, M. P., McGee, K. A., et al. (2001). Soil efflux and total emission rates of magmatic CO_2 at the Horseshoe lake tree kill, Mammoth mountain, CA, 1995–1999. *Chemical Geology, 177,* 101–116.
71. Cardellini, C., Chiodini, G., Frondini, F., et al. (2003). Accumulation chamber measurements of methane fluxes: Application to volcanic-geothermal areas and landfills. *Applied Geochemistry, 18,* 45–54.
72. Tassi, F., Montegrossi, G., Vaselli, O., et al. (2009). Flux measurements of benzene and toluene from landfill cover soils. *Waste Management and Research, 29,* 50–58.
73. Mazot, A., & Taran, Y. (2009). CO_2 flux from the volcanic lake of El Chichón (Mexico). *Geofísica Internacional, 48,* 73–83.
74. Sinclair, A. J. (1974). Selection of threshold values in geochemical data using probability graphs. *Journal of Geochemical Exploration, 3,* 129–149. https://doi.org/10.1016/0375-6742(74)90030-2
75. Sichel, H. S. (1966). The estimation of means and associated confidence limits for smalls samples from lognormal populations. In *Symposium on Mathematical Statistics and Computer Applications in Ore Valuation* (pp. 106–122). South African Institute of Mining and Metallurgy.
76. Deutsch, C. V., & Journel, A. G. (1998). *GSLIB: Geostatistical software library and users guide* (2nd ed.). New York: Oxford University Press.
77. Metcalf, A. E. (2014). *Trazabilidad isotópica del carbono: Implicaciones en el almacenamiento geológico de CO_2* (PhD thesis). Universidad Internacional Menéndez Pelayo, 233 pp.
78. Vogel, J. C., Grootes, P. M., & Mook, W. G. (1970). Isotopic fractionation between gaseous and dissolved carbon dioxide. *Zeitschrift für Physik, 230,* 225–238.
79. Deines, P. (1980). The isotopic composition of reduced carbon. In A. Fritz & P. Fontes (Eds.), *The terrestrial environment. Handbook of environmental isotope geochemistry* (pp. 329–434). Elsevier Scientific Press.
80. O'Leary, M. H. (1988). Carbon isotopes in photosynthesis. *BioScience, 38,* 328–336.
81. Friedli, H., Lotscher, H., Oeschger, H., et al. (1986). Ice core record of the $^{13}C/^{12}C$ ratio of atmospheric CO_2 in the past two centuries. *Nature, 324,* 237–238.
82. Javoy, M., Pineau, F., & Delorme, H. (1986). Carbon and nitrogen isotopes in the mantle. *Chemical Geology, 57,* 41–62.
83. Clark, I. D., & Fritz, P. (1997). *Environmental isotopes in hydrogeology* (p. 328). New York: CRC Press.
84. Cerling, T. E. (1984). The stable isotopic composition of modern soil carbonate and its relationship to climate. *Earth and Planetary Science Letters, 71,* 229–240.
85. Cerling, T. E. (1991). Carbon dioxide in the atmosphere: Evidence from Cenozoic and Mesozoic paleosols. *American Journal of Science, 291,* 377–400.
86. Romanek, C. S., Grossman, E. L., & Morse, J. W. (1992). Carbon isotopic fractionation in synthetic aragonite and calcite: Effects of temperature and precipitation rate. *Geochimica et Cosmochimica Acta, 56,* 419–430.

87. Reyes, E., Pérez del Villar, L., Delgado, A., et al. (1998). Carbonatation processes at the El Berrocal analogue granitic system (Spain): Mineralogical and isotopic study. *Chemical Geology, 150*, 293–315.
88. Jenkins, A. C., & Cook, A. (1961). *Argon, helium and the rare gases: History, occurrence and properties*. London: Interscience Publishers.
89. Clarke, W. B., Beg, M. A., & Craig, H. (1969). Excess ^3He in the sea: Evidence for terrestrial primordial helium. *Earth and Planetary Science Letters, 6*, 213–220.
90. Sano, Y., & Wakita, H. (1985). Geographical distribution of $^3He/^4He$ ratios in Japan: Implication for arc tectonics and incipient magmatism. *Journal of Geophysical Research, 90*, 8719–8741.
91. Oxburgh, E. R., O'Nions, R. K., & Hill, R. I. (1986). Helium isotopes in sedimentary basins. *Nature, 324*, 632–635.
92. Hiyagon, H., & Kennedy, B. M. (1992). Noble gases in CH_4-rich gas fields, Alberta, Canada. *Geochimica et Cosmochimica Acta, 56*, 1569–1589.
93. Craig, H., & Lupton, J. E. (1976). Primordial neon, helium, and hydrogen in oceanic basalts. *Earth and Planetary Science Letters, 31*, 369–385.
94. Kurz, M. D., & Jenkins, W. J. (1981). The distribution of helium in oceanic basalt glasses. *Earth and Planetary Science Letters, 53*, 41–54.
95. Lupton, J. E. (1983). Terrestrial inert gases: Isotope tracer studies and clues to primordial components in the mantle. *Annual Review of Earth and Planetary Sciences, 11*, 371–414.
96. Ozima, M., & Zashu, S. (1983). Noble gases in submarine pillow volcanic glasses. *Earth and Planetary Science Letters, 62*, 24–40.
97. Kaneoka, I., & Takaoka, N. (1980). Rare gas isotopes in Hawaiian ultramafic nodules and volcanic rocks: Constraints on genetic relationships. *Science, 208*, 1366–1368.
98. Marty, B., Meynier, V., Nicolini, E., et al. (1993). Geochemistry of gas emanations: A case study of the Réunion Hot Spot, Indian Ocean. *Applied Geochemistry, 8*, 141–152.
99. Ozima, M., & Podosek, F. A. (2002). *Noble gas geochemistry* (2nd ed., 286 pp.). Cambridge University Press.
100. Sano, Y., Takahata, N., Nishio, Y., et al. (1998). Nitrogen recycling in subduction zones. *Geophysical Research Letters, 25*, 2289–2292.
101. Fischer, T. P., Hilton, D. R., Zimmer, M. M., et al. (2002). Subduction and recycling of nitrogen along the Central American margin. *Science, 297*, 1154–1157.
102. Werner, R. A., & Brand, W. A. (2001). Referencing strategies and techniques in stable isotope ratio analysis. *Rapid Communications in Mass Spectrometry, 15*, 501–519.
103. Neele, F., Grimstad, A.-A., Fleury, M., et al. (2014). MiReCOL: Developing corrective measures for CO_2 storage. *Energy Procedia, 63*, 4658–4665.
104. Johnson, J. W., Nitao, J. J., Morris, J. P., et al. (2003). Reactive transport modeling of geohazards associated with CO_2 injection for EOR and geologic sequestration. In *Offshore Technology Conference, OTC 15119*.
105. Lavrov, A. (2016). Dynamics of stresses and fractures in reservoir and cap rock under production and injection. *Energy Procedia, 86*, 381–390.
106. Antropov, A., Lavrov, A., & Orlic, B. (2017). Effect of in-situ stress alterations on flow through faults and fractures in the cap rock. *Energy Procedia, 114*, 3193–3201.
107. Maldonado, R. (2017). *Riesgos geomecánicos asociados a la inyección de CO_2 en formaciones geológicas* (PhD thesis). Universidad de Oviedo.
108. Durucan, S., Korre, A., Shi, J.-Q., et al. (2016). The use of polymer-gel solutions for CO_2 flow diversion and mobility control within storage sites. *Energy Procedia, 86*, 450–459.
109. Wasch, L. J., Wollenweber, J., Neele, F., et al. (2017). Mitigating CO_2 leakage by immobilizing CO_2 into solid reaction products. *Energy Procedia, 114*, 4214–4226.
110. Govindan, R., Si, G., Korre, A., et al. (2017). The assessment of CO_2 backproduction as a technique for potential leakage remediation at the Ketzin pilot site in Germany. *Energy Procedia, 114*, 4154–4163.
111. Bossie-Codreanu, D. (2017). Study of N_2 as a mean to improve CO_2 storage safety. *Energy Procedia, 114*, 5479–5499.

112. MiReCOL. (2017). *Final report summary—MiReCOL (Mitigation and Remediation of CO_2 leakage)*. https://cordis.europa.eu/project/id/608608/reporting. Accessed November 12, 2020.
113. Orlic, B., Loeve, D., & Peters, E. (2016). *Remediation of leakage by diversion of $CO_2$to nearby reservoir compartments*. MiReCOL Publications. https://www.mirecol-co2.eu/publications.html Accessed June 24, 2020.
114. Mosleh, M. H., Govindan, R., Shi, J.-Q., et al. (2017). The use of polymer-gel remediation for CO_2 leakage through faults and fractures in the caprock. *Energy Procedia, 114*, 4164–4171.
115. Réveillère, A., & Rohmer, J. (2011). Managing the risk of CO_2 leakage from deep saline aquifer reservoirs through the creation of a hydraulic barrier. *Energy Procedia, 4*, 3187–3194.
116. Benson, S. M. (2006). *Monitoring carbon dioxide sequestration in deep geological formations for inventory verification and carbon credits* (SPE 102833). Society of Petroleum Engineers.
117. Celia, M. A., Bachu, S., Nordbotten, J. M., et al. (2004). Quantitative estimation of CO_2 leakage from geological storage: Analytical models, numerical models and data needs. In *Greenhouse gas control technologies* (pp. 663–671). Oxford: Elsevier Science Ltd.
118. Fleury, M., Sissmann, O., Brosse, E., et al. (2017). A silicate based process for plugging the near well bore formation. *Energy Procedia, 114*, 4172–4187.
119. Manceau, J.-C., Hatzignatiou, D. G., de Latour, L., et al. (2014). Mitigation and remediation technologies and practices in case of undesired migration of CO_2 from a geological storage unit—Current status. *International Journal of Greenhouse Gas Control, 22*, 272–290.
120. Mosleh, M. H., Durucan, S., Syed, A., et al. (2017). Development and characterisation of a smart cement for CO_2 leakage remediation at wellbores. *Energy Procedia, 114*, 4154–4163.
121. Todorovic, J., Raphaug, M., Lindeberg, E., et al. (2016). Remediation of leakage through annular cement using a polymer resin: A laboratory study. *Energy Procedia, 86*, 442–449.
122. Pearl, J. (2000). *Causality. Models, reasoning and inference*. Nueva York: Cambridge University Press.
123. Gigerenzer, G. (1996). On narrow norms and vague heuristic: A reply to Kahneman and Tversky. *Psychological Review, 103*, 592–596.
124. Seldmeier, P., & Gigerenzer, G. (2001). Teaching bayesian reasoning in less than two hours. *Journal of Experimental Psychology General, 130*, 380–400.

Risk Communication and Stakeholder Engagement in CCS Projects. Lessons Learned from the Compostilla Project

Jordi Bruno and Juan Castaño

Abstract In this Chapter we review the main social and cultural factors that have impaired the development of the CCS technology in Spain and elsewhere, with a special focus on the hurdles encountered in the development of the Compostilla project and the actions that were taken to unlock the situation at a local scale. Honest and trustworthy scientific information and open dialogue were the key factors for success.

Keywords CCS · Social acceptance · Compostilla project · Ciuden

1 Introduction

The matter of social acceptance and risk perception of energy related technologies and projects has been the objective of our work for more than 20 years. We developed our methodology initially because of the hurdles we faced in implementing our scientific and technological solutions in the field of nuclear waste management.

Although, we are a scientific and technological company, we realised quite early that in order to develop useful solutions to sort environmental challenges, we had to open a dialogue with the concerned societies to understand and address their legitimate concerns.

Already at the onset of the CCS projects, it was clear to us that one of the key challenges in the implementation of the geological storage technology would be its social perception and acceptance.

In one of the first EU workshops organised in 2006 in Austria, we pointed out that based on our experience on nuclear waste management (NWM) programmes, the CCS implementation would face similar challenges for its social acceptance. The opinion from the CCS community at this point was that CCS and NWM had very little in common, that they did not want to get mixed with this "evil" technology

J. Bruno (✉) · J. Castaño
Amphos 21 Group, Barcelona, Spain
e-mail: jordi.bruno@amphos21.com

J. C. de Dios et al. (eds.), *CO_2 Injection in the Network of Carbonate Fractures*, Petroleum Engineering, https://doi.org/10.1007/978-3-030-62986-1_8

and that CCS had a very strong social argument as a technology for climate change mitigation.

Many of the key actors at that point in time had very high expectations about the implementation of the CCS technology in the context of climate change and there was an explosion of projects supported by the EC with very large investments involved. Some energy companies were also attracted by this financial pool and participated in these projects, particularly Shell, Vatenfall, PGE and Endesa.

Fourteen years later, we have seen that these arguments were quite naïve and that many CCS programmes have been stalled due to social perception and acceptance problems and the enormous economic investment has been basically gone in combination with the failure of the carbon market as such.

The social failure on the implementation of the CCS projects based on the global benefit of climate change mitigation indicated that many other factors have to be considered. This has triggered a substantial body of literature in the social sciences community which has contributed to frame some of the key factors concerning the public acceptance or rejection of the CCS technology. The reader is referred to several review works that have been published between 2012 and 2016 which cover most of the social research conducted around the public an risk perception of the CCS technology [2–4].

This chapter does not pretend to add or to substitute this literature rather our objective is to frame some of these concepts in the real application and experiences we have been subjected to in our work. Particularly, in the context of the Compostilla project and the sitting of the Hontomin Pilot Injection Plant, led by Ciuden and the investigations of a potential injection site in Tierra de Campos in the Castilla y Leon region, promoted by Endesa.

Most of the social research work quoted in these articles is mainly based on the analysis of interviews to some specific population samples, normally quite limited in size and therefore in representativity. However, they constitute the base for some of the key opinions and trends that develop around this kind of projects.

The largest pool of data regarding attitudes towards CCS implementation was performed in 2011 by the Eurobarometer [1].

The data collected indicated some interesting variability depending on the cultural values of the various countries considered and some valuable conclusions were extracted by Karini et al. [2].

Since, the focus of this work is the implementation of CCS in Spain, we will concentrate on the results of this Eurobarometer survey in Spain compared to the EU average. The time of the survey corresponds to the actual period when our work concerning the social acceptance of the Compostilla project was in place and therefore it constitutes a representative framework.

As we already mentioned, the CCS technology was been promoted in the context of climate change abatement, it is interesting to see how the various countries responded to the question related to climate change awareness and knowledge (Fig. 1).

As it can be seen, in Spain a 59% of the population felt uninformed about the consequences of climate change by 2011, against an average of 51% in the overall

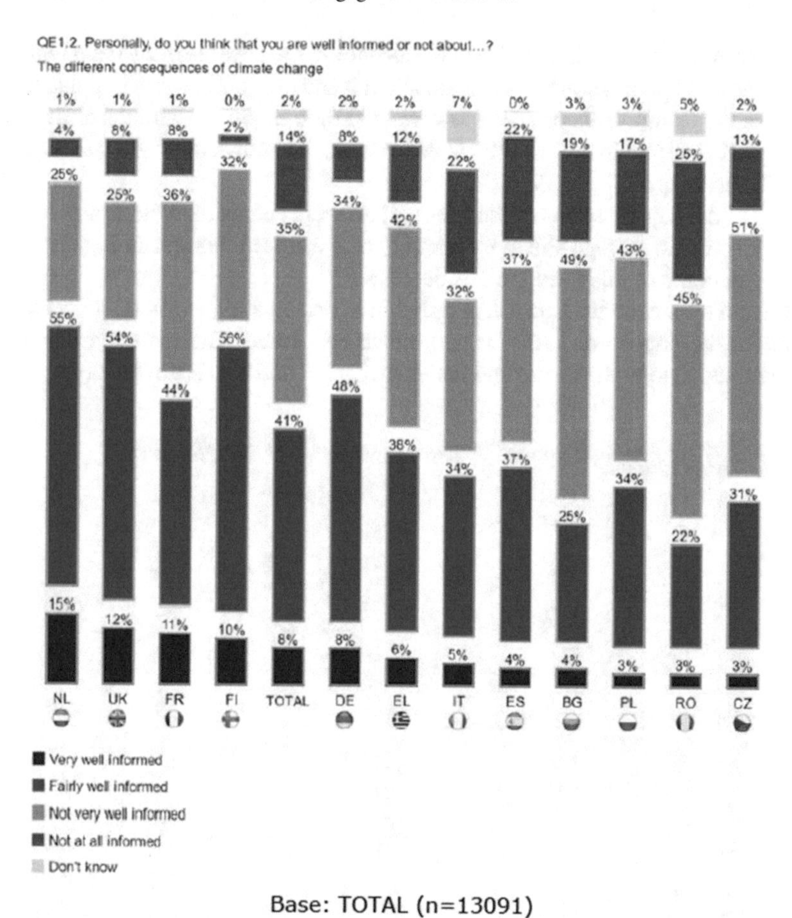

Fig. 1 Results from Eurobarometer [1] concerning the awareness on the consequences of climate change

EU. We have no reason to believe that the balance has improved in the last decade. As the authors of the Eurobarometer indicated [1], there is a strong direct correlation between the level of education and the knowledge and awareness about climate change issues and these has notorious consequences in Spain, a country with one of the largest early school dropout indexes in basic education.

This is also visible in the answers concerning the knowledge about CO_2 and its consequences, which show an appalling lack of basic understanding, although up to 88% of the Spanish respondents indicated that CO_2 had a large impact on climate change. This is an interesting answer from a population that had a low knowledge ratio on climate change and carbon dioxide, probably due to the fact that the information channels of the Spanish population at this time was television and this connection was duly amplified in the main television channels.

Concerning the knowledge of the CCS technology, the results from Spain indicated that 83% of the population had no heard about it and only a 2% of the population was aware of the Compostilla project. However, the ratio improved with the closeness to the project sites and in the Castilla y Leon region, the knowledge ratio was higher as we will see later on.

One of the key questions of the Eurobarometer [1] concerned the risk perception of placing a CO_2 storage site in the vicinity of the current homes. In Fig. 2, we can see the outcome of this question by countries.

The answers from the Spanish population indicated a high level of concern, with a 67% of the population fairly or very much concerned. The main reasons being, potential damage to the health and the environment and the risk of leakage.

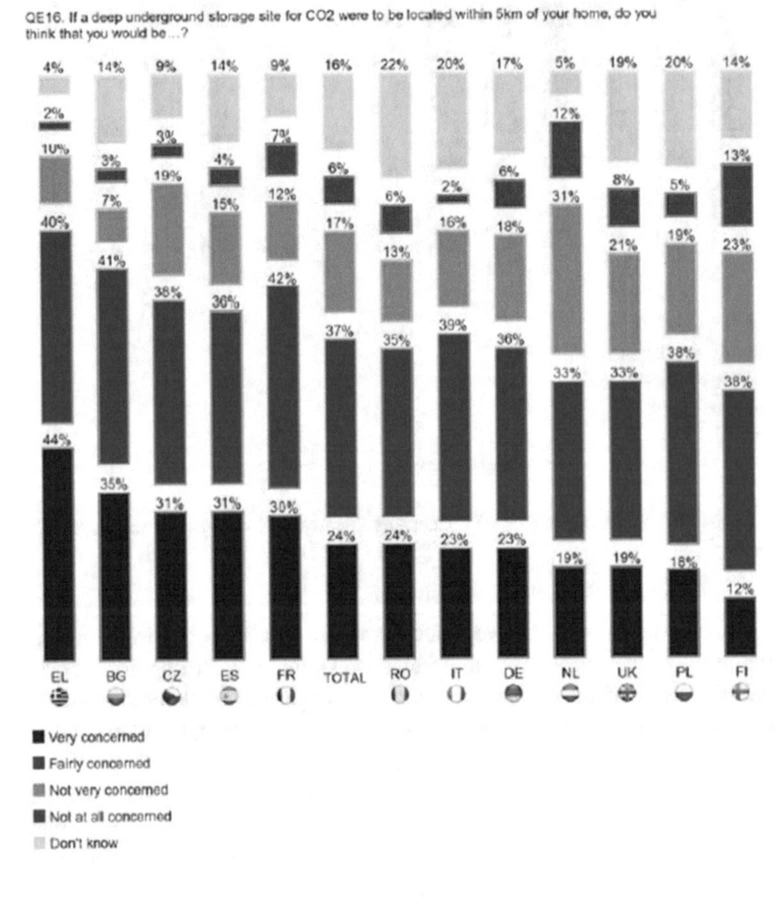

Fig. 2 Eurobarometer [1] answers to the question concerning the sitting of CO_2 storage in the vicinity (5 km) of the homes of the consulted population

One of the positive outcomes of the Eurobarometer study was that the majority of the population wanted to be involved in the decision-making process. This is, in our opinion one of the fundamental aspects to be taken into consideration, as we will develop later on.

It was this social context that we performed out strategies for social and risk perception of the Spanish CCS projects, particularly the ones related to the Compostilla oxy-combustion project.

What follows is an account of our experience when dealing with the social acceptance of CCS projects, particularly the ones that have been launched in Spain. The objective is to bring about a roadmap to prevent, or at least to mitigate, the social impact in future carbon storage projects.

2 Lessons from the Experiences Around the Compostilla Project

We already learned from our experience in nuclear waste management projects that the social analysis and communication efforts should be implemented at the onset of the project. The earlier this is done, the better is the social understanding and acceptance of a project which is constructed together, rather than imposed and then informed about it.

The communication strategy varies depending on the different levels, but it should focus on the local area, as well as in the regional perspective. This is particularly important in Spain, where there are three/four administration levels (municipal, regional/county councils and the national government). In many cases, the political differences among the various administrations use environmental projects as battlefields and therefore a large consensus should be aimed to in order to guarantee the success of the project.

This means that strong support on the local area (the community) does not guarantee support at the higher political level as it was evidenced during the sitting process of the temporary storage facility for spent nuclear fuel. Although, the support for sitting the facility in Ascó (Spain) was very strong at the local scale, the regional Catalan government vetoed the project for short-term political reasons.

In spite of these difficulties it is clear to us that there is a need to have a basic methodology to approach the dialogue and engagement of the local communities in technological projects and in particular in the ones related to CCS implementation. This methodology has been developed throughout the work of the Amphos 21 team in many environmental projects which involved some social controversy.

3 The Methodology

The development of the social and risk perception analysis has three main components

Social characterization
Communication and implication
Social monitoring

The social characterization component is a fundamental part of the methodology, it is very important to understand which are the main key drivers in risk perception and who are the key actors in the development of the risk perception and opinion. The outcome of this first step is the basis for planning and developing the communication and social implication actions of the second step.

The communication and implication actions to be developed are very dependent on the outcome of the social characterization and of the various levels to be approached, local, regional, national or international.

- At the international level the scientific prestige of the project is fundamental, and this is achieved by participating in international fora and on scientific publications.
- At the national level, there had to be an effort to inform about the technology, this is particularly important in Spain. As we have seen, the level of knowledge is rather low.
- At the regional level, the efforts on the information about the technology have to be strong. As we have already seen, the efforts payed of, since the knowledge about CCS was comparatively higher in Castilla y Leon (the region were the Compostilla project sites were located) than in the rest of Spain.
- At the local level is where the communication efforts have to be stronger in order to involve the local community in the development of the project and the potential outcomes for the development of the territory in order to build common trust.

This is particularly visible in the outcome of the social characterization study that Ciuden promoted in 2011 in the local area around the Compostilla and Hontomin sites.

The outcome of the study indicated that the local awareness increased around the two sites. While the average level of awareness of the population in the Castilla y Leon region showed to be close to 15% (percentage of the population that knew about the project), the level of awareness doubled in the Burgos province (28%) and was up to 50% in the Leon province. This was the result of a very clear strategy of local and regional communication and involvement promoted by Ciuden, as pointed out in Oltra et al. [4]. The fact that the local communication and involvement was led by one of the key scientists of the project, Andrés Pérez Estaún from CSIC, was a fundamental factor for the success of the project.

The communication and involvement actions have to be continuous through the various phases of the project and very flexible to respond to the various concerns they may arise from the various stakeholders.

The third step of the methodology consists on the monitoring and evaluation of the communication and involvement actions and a contingency plan should be devised for unexpected events.

4 Tierra de Campos, Sahagun (Leon), a Case Study

August 8 of 2011, on a Sunday evening we received a call from the manager of the Compostilla project at Endesa. The reason was that the development of their 3D seismology programme in Tierra de Campos, one of the sites they were investigating for geological storage, had been stopped by the local population and they required the help of Amphos 21 to unlock the situation.

The social context of Tierra de Campos is basically rural, with a very low population density and the local communities are very aged.

The site investigations had been started by 2010 and that triggered some interest in the area as there were speculations about the existence of oil and gas resources in the area. This was common also to the Hontomin site, because there had been some oil explorations in the 60s around the area.

This generated a lot of rumours and expectations due to the lack of clear information. By 2011 some information had only been provided to the majors of the villages involved and to the owners of the land concerned in the investigations.

One of the main problems was that the deep borehole construction had been started without any information to the local population. This lack of information triggered rumours about the use of the site to dump nuclear waste and some community platforms were created against the project.

The opposition to the project crystallised in fact that the permit for the 3D seismic campaign was denied by the local authorities pressured by the local population and the landowners.

To turn around the situation we applied the communication and risk analysis methodology previously developed in previous energy related projects.

We started by applying the first step of the methodology, to analyse the local social situation. The result of the analysis indicated several key issues:

1. Lack of information about the purpose of the project
2. Risk perception concerning the potential leakage of CO_2
3. Risk of explosion, a link to the Fukushima accident was repeatedly mentioned
4. Risk of groundwater contamination
5. Absence of benefits to the local citizens, basically the farmers that were affected by the deployment of the 3D seismology work.

The stakeholder analysis of the municipalities involved gave the result depicted in Fig. 3.

Basically, most of the municipalities and the local newspapers had an ambiguous position towards the project and there was a specific municipality that was against it.

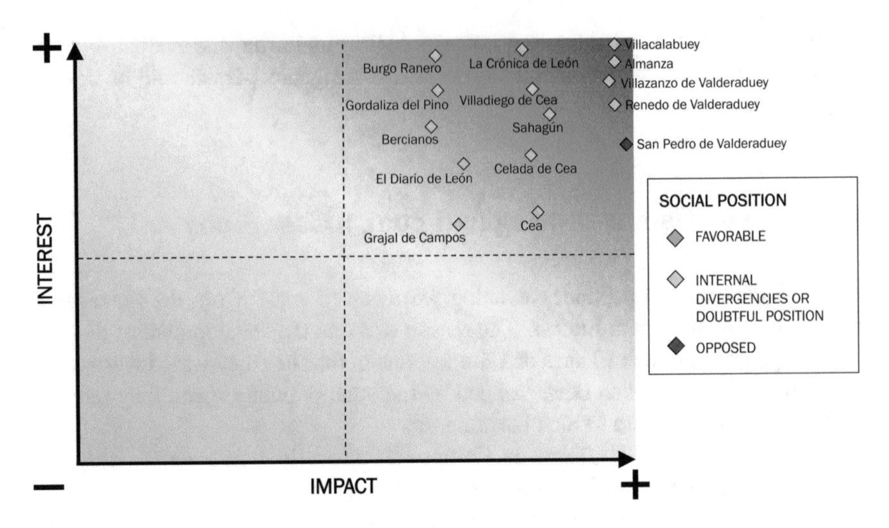

Fig. 3 Result of the stakeholder analysis performed by Amphos 21

Certainly, the situation required a very proactive communication and involvement strategy, which we run intensively in an eight-week period. Some of the actions were performed with the support of Ciuden which was visualised by the locals as a trustworthy institution led by scientists [4]. The main actions implemented were:

- We run an information meeting with all the majors involved as a prelude to 8 public information meetings to the local population of the various villages involved, focused on explaining the scope of the investigations in the context of CCS and climate change abatement.
- Public informative visits were organised to the borehole to explain the purpose of the deep borehole investigations.
- Ciuden organised visits of the local citizens and schools to the CO_2 capture plant in Compostilla.
- There was a permanent effort of information to the local press with visits to the borehole site and continuous press releases.
- A stand was placed in the Local Fair of the County in Sahagun, to inform about the scope of the project.
- A continuous social monitoring effort was deployed to have a continuous communication with the stakeholders.

The outcome of this effort was successful, and the site investigations were continued until Endesa decided to finalise the project by 2013.

5 Conclusions

The implementation of CCS has faced many hurdles, many of them have political and economical dimensions which are beyond the scope of this analysis. However, one of the key bottlenecks is its social acceptance in the frame of the risk perception of the technology.

As a matter of fact, the negative social perception of the CCS implementation in some of the sites, triggered the abandon of the projects by the energy companies that were securing the investment.

The implementation of any technological project requires a strong commitment on transparency, information and public engagement from the onset.

Risk perception is a complex psychological and social issue and requires to be dealt at many different scales. The role of scientists and technologists in this area is very important to secure that the information is rigorous and appropriated.

In terms of CCS implementation, as pointed out by Oltra et al. [4], the projects led by scientists have a stronger credibility than the ones led by energy companies. This was evident in the different approach used in Hontomin versus Tierra de Campos, within the Compostilla project.

However, a dedicated information and engagement work with the local communities in Tierra de Campos made possible to turn around a deadlock situation and the project could continue with the support of the local communities.

Acknowledgements The experiences and results showed in this chapter form part of the project "OXYCFB 300, Compostilla" funded by the European Energy Program for Recovery (EEPR). Authors acknowledge the role of the funding entities, project partners and collaborators without which the project would not have been completed successfully.

Jordi Bruno wants to acknowledge the fundamental contribution of the late Andrés Pérez Estaún a superb geologist, a magnificent communicator and an even better person. He is very much missed!

References

1. Eurobarometer. (2011). *Special Eurobarometer 364. Public acceptance and awareness of CO_2storage. DG energy and DG communication.* European Commission.
2. Karini, F., Toikka, A., & Hukkinen, J. I. (2016). Comparative socio-cultural analysis of risk perception of carbon capture and storage in the European Union. *Energy Research and Social Science, 21,* 114–122.
3. L'Orange Seigo, S., Dohle, S., & Siegrist, M. (2014). Public perception of carbon capture and storage (CCS): A review. *Renewable and Sustainable Reviews, 38,* 844–863.
4. Oltra, C., Upham, P., Riesch, H., Boso, A., Brusting, S., Dütschke, E., & Lis, A. (2012). Public responses to CO_2 storage sites: Lessons from five European cases. *Energy & Environment, 23*(2), 227–248.

Printed in the United States
by Baker & Taylor Publisher Services